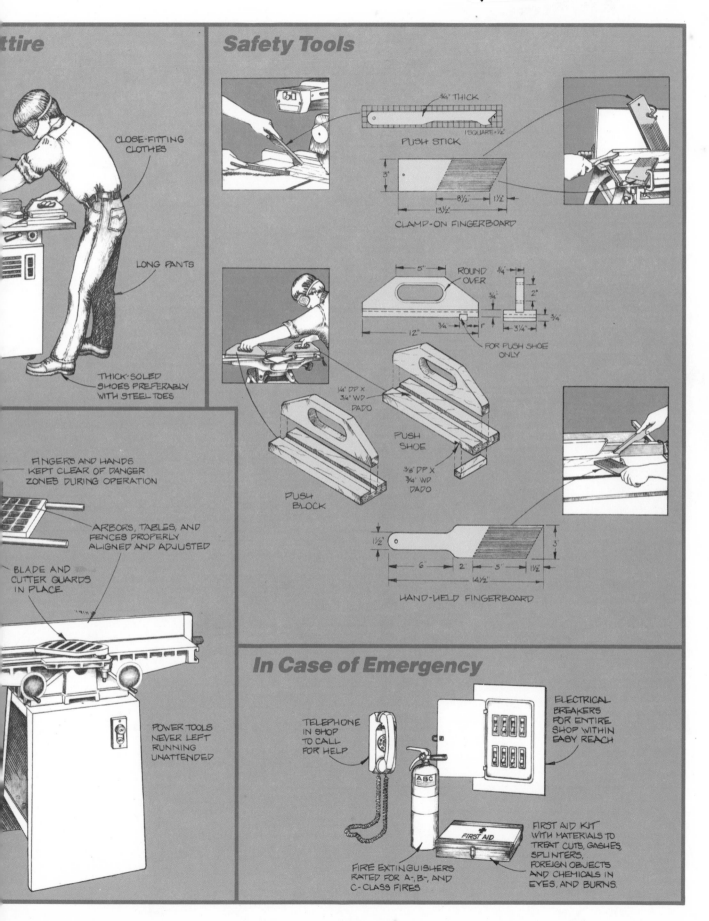

ttire

CLOSE-FITTING CLOTHES

LONG PANTS

THICK-SOLED SHOES PREFERABLY WITH STEEL TOES

FINGERS AND HANDS KEPT CLEAR OF DANGER ZONES DURING OPERATION

ARBORS, TABLES, AND FENCES PROPERLY ALIGNED AND ADJUSTED

BLADE AND CUTTER GUARDS IN PLACE

POWER TOOLS NEVER LEFT RUNNING UNATTENDED

Safety Tools

¾" THICK

PUSH STICK

1 SQUARE = ½"

CLAMP-ON FINGERBOARD

3"

8½"

1½"

13½"

5"

ROUND OVER

¾"

¾"

2"

¾"

12"

¾"

1"

3¼"

FOR PUSH SHOE ONLY

¼" DP X ¾" WD DADO

PUSH SHOE

⅜" DP X ¾" WD DADO

PUSH BLOCK

1½"

6"

2"

5"

1½"

3"

14½"

HAND-HELD FINGERBOARD

In Case of Emergency

TELEPHONE IN SHOP TO CALL FOR HELP

ELECTRICAL BREAKERS FOR ENTIRE SHOP WITHIN EASY REACH

ABC

FIRST AID

FIRE EXTINGUISHERS RATED FOR A-, B-, AND C-CLASS FIRES

FIRST AID KIT WITH MATERIALS TO TREAT CUTS, GASHES, SPLINTERS, FOREIGN OBJECTS AND CHEMICALS IN EYES, AND BURNS.

·BUILD·IT·BETTER·YOURSELF·
WOODWORKING PROJECTS

Desks and Bookcases

Collected and Written
by Nick Engler

Rodale Press
Emmaus, Pennsylvania

If you have any questions or comments concerning this book, please write:
    Rodale Press
    Book Reader Service
    33 East Minor Street
    Emmaus, PA 18098

Series Editor: Jeff Day
Managing Editor/Author: Nick Engler
Editor: Roger Yepsen
Copy Editor: Mary Green
Graphic Designer: Linda Watts
Graphic Artists: Mary Jane Favorite
                Chris Walendzak
Photography: Karen Callahan
Cover Photography: Mitch Mandel
Cover Photograph Stylist: Janet C. Vera
Proofreader: Hue Park
Typesetting by Computer Typography, Huber Heights, Ohio
Interior Illustrations by O'Neil & Associates, Dayton, Ohio
Endpaper Illustrations by Mary Jane Favorite
Produced by Bookworks, Inc., West Milton, Ohio

Library of Congress Cataloging-in-Publication Data

Engler, Nick.
    Desks and bookcases/collected and written by Nick
Engler.
        p.   cm.—(Build-it-better-yourself
        woodworking projects)
    ISBN 0–87857-·847–1 hardcover
    ISBN 0–87857–848–X paperback
    1. Furniture making. 2. Desks.
    3. Bookcases.   I. Title.
    II. Series: Engler, Nick. Build-it-better-yourself
woodworking projects.
TT197.5.D4E54   1989
684.1′6—dc20   89–7028
                                                      CIP

Distributed in the book trade by St. Martin's Press

2   4   6   8   10   9   7   5   3   hardcover
2   4   6   8   10   9   7   5   3   1   paperback

Contents

Paperwork

Like so many modern inventions, we've come to think of desks and bookcases as necessities. After all, if you didn't have a desk, where would you pay your bills? And without bookcases, how would you store and organize the hundreds of books, magazines, and other publications that we depend on?

These weren't always problems, however. As furniture designs go, desks and bookcases are relative newcomers, spawned by the burgeoning paperwork of modern civilization — letters, bills, taxes, catalogs, magazines, and so on. This paperwork is a by-product of another new development — literacy.

Before the mid-eighteenth century, the vast majority of people in America and Europe were illiterate. Our ancestors had no need for reading and writing; all they needed to know was handed down from their ancestors. They either worked the land or practiced a craft. Their lives were stable and changeless.

Their furniture reflected this. Homes contained simple tables, benches, beds, and chests. With few exceptions, there was no place to write a letter or store a book. If a family member was ambitious enough to learn to read, it was usually to study the Bible. This single book was kept in a small chest called a *Bible box*. If that person could also write, they might have had a *lap desk*. This was similar to a Bible box, but it had a slanted lid that served as a writing surface. When in use, it sat upon a person's lap. Both these pieces were luxuries for common folks.

The Industrial Revolution changed this bucolic existence forever. As new ideas, tools, methods, and products were introduced, new ways of working and living began to evolve. This evolution was fueled by an exchange of information. Folks learned of an innovation, added their own refinements, and passed them along. Because the information was mostly written, people learned to read and write to keep up with the progress. A public school system developed to teach them, and for the first time in history, literacy became the rule rather than the exception.

In the span of a few generations, people began to correspond, keep records, write contracts, and read all kinds of books. To handle the paperwork, cabinetmakers made larger lap desks with more storage for paper, pens, and ink. As these desks grew beyond lap size, they built special tables to put them on. Before long, an innovative craftsman attached the legs directly to the case to make the first stand-alone desk.

The bookcase evolved in a different manner. The automation of the printing press and book binding in the early nineteenth century suddenly made books plentiful and cheap. In a short time, home libraries expanded from one book to dozens. Books moved out of Bible boxes and into hutches, cupboards, anything big enough to hold them. If these shelving units weren't enclosed already, doors were added to protect the books from insects and rodents. Bible boxes faded away, an evolutionary dead-end in the history of furniture. In their stead, the converted shelves became the first bookcases.

As literacy advanced, so did the accompanying paperwork. No longer luxuries, desks and bookcases became necessities to deal with the chores of a literate life. Many different types of desks and bookcases evolved, tailored to specific types of paperwork — secretaries, writing tables, rolltop desks, expandable and adjustable bookcases, to name a few.

By the turn of the century, the ever-increasing paperwork was spilling out of the desks and bookcases and into even newer furniture forms — filing cabinets, taborets, drafting tables, and so on.

In recent years, the Industrial Revolution has given way to an electronic revolution. A new type of literacy has developed — computer literacy. As we're swept along by yet another wave of paperwork — video paperwork — desk and bookcase designs are evolving again. The phenomenal growth of computers has already spawned a whole new line of office furniture and accessories, such as computer workstations and diskette files. These, too, are becoming necessities as computers grow more and more pervasive in everyday life.

Contemporary/Classic Writing Table

During the late eighteenth century, when the classic cabinetmaking shops of Newport and Philadelphia were at their height, many new types of furniture developed. Among them was the *writing table*. It was a small, rectangular table, just wide enough for a person to write comfortably, but long enough to keep a lamp, an inkwell, and reference materials on the tabletop. A single drawer in the middle of the table stored extra paper, pens, and nibs.

The table shown follows this classic formula, with a few contemporary twists. It was built by W. R. Goehring, an accomplished cabinetmaker from Gambier, Ohio. He borrowed the overall design from American Chippendale woodworking. The curve of the legs, however, is country German or Moravian, rather than the French and Italian cabrioles normally found on Chippendale pieces. He carved the drawer pull from wood and inlaid it in the drawer front — a feature common to contemporary woodworking. Yet the shape of the pull suggests the brass shield-and-bail used during the Federal period.

"I interpret the classic and country forms," says Goehring, "and blend them into a modern piece. My woodworking has historical roots, but a contemporary look."

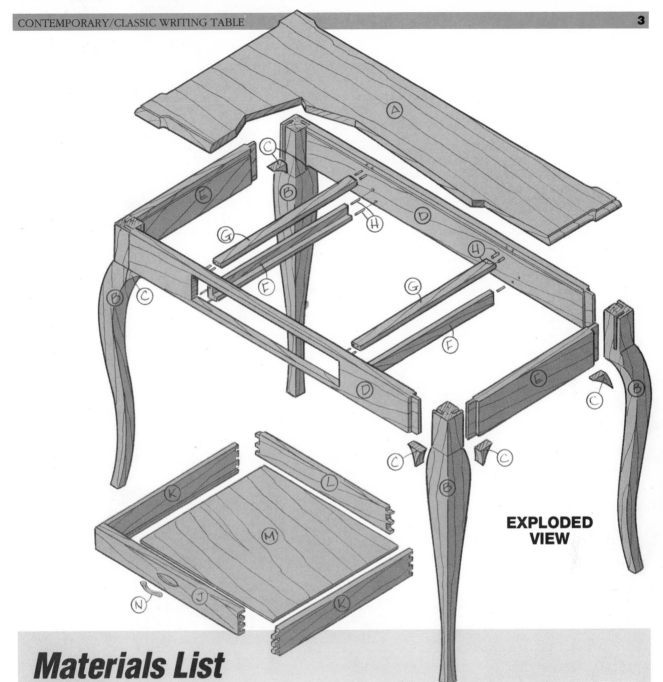

EXPLODED VIEW

Materials List

FINISHED DIMENSIONS

PARTS

A. Tabletop   ¾″ x 21″ x 42″

B. Legs (4)   2¾″ x 2¾″ x 28¼″

C. Ears (8)   1¼″ x 1½″ x 1¾″

D. Front/back aprons (2)   ¾″ x 4″ x 38″

E. Side aprons (2)   ¾″ x 4″ x 17″

F. Drawer guides (2)   1″ x 1⅛″ x 17½″

G. Kickers (2)   ½″ x 1″ x 17½″

H. Dowels (16)   ¼″ dia. x 1″

J. Drawer front   ¾″ x 2¹⁵⁄₁₆″ x 17⅞″

K. Drawer sides (2)   ½″ x 2¹⁵⁄₁₆″ x 17″

L. Drawer back   ½″ x 2⁷⁄₁₆″ x 17⅞″

M. Drawer bottom   ¼″ x 16¹⁵⁄₁₆″ x 17⅜″

N. Drawer pull   ⁷⁄₁₆″ x ½″ x 3″

HARDWARE

Tabletop clips (8)

#6 x ⅝″ Flathead wood screws (9)

1

Select and plane the stock. To make this project, you'll need approximately 12 board feet of 4/4 (four-quarters) stock, 10 board feet of 12/4 (twelve-quarters) stock, and one-quarter of a 4' x 8' sheet of cabinet-grade ¼" plywood. Most classic writing desks were made from mahogany, although some cabinet-makers used walnut and cherry. You can save a little money by using an inexpensive wood for the parts that don't show — drawer sides, back, guides, and kickers.

In the table shown, the primary wood is walnut, and the secondary is poplar.

Rip the leg pieces from 12/4 stock and set them aside. Then plane stock for the ears and drawer guides from the remaining 12/4 wood. Plane the remaining stock to the thicknesses needed. Use the 4/4 stock for the top, aprons, kickers, and drawer parts (except the bottom).

Don't cut any of the parts to size yet. On this particular project, it's easier to cut them as you go.

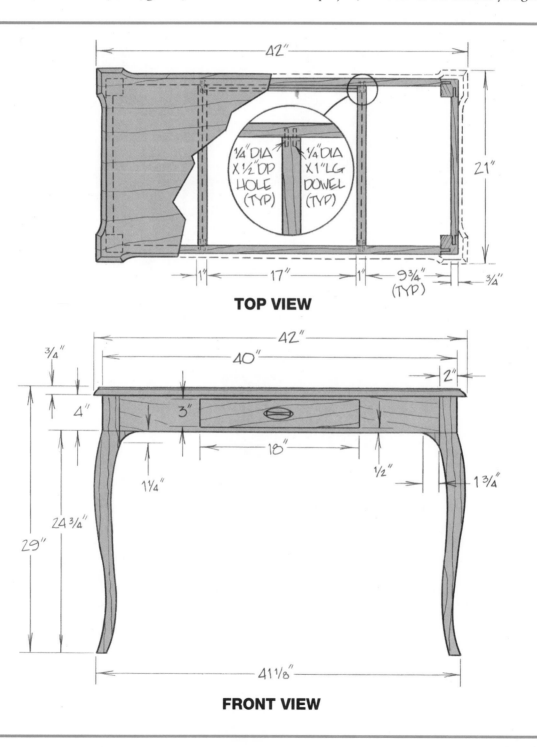

TOP VIEW

FRONT VIEW

2 Square the legs and cut the joinery.

Square the leg stock on a jointer, and plane it to the width and thickness needed. (See Figures 1 through 3.) Cut the four legs to length, and cut the ear blocks to size.

To mark the leg shape on the stock, first make a template. Enlarge the *Leg and Ear Pattern* and trace it onto a scrap of plywood or hardboard. Mark a horizontal line across the leg shape where the straight portion (or *post*) of the leg ends and the curved portion (or *knee*) begins. This line will help you to align the template on the stock. Cut out the template with a band saw and sand the edges smooth.

2/Turn the wood 90°, so the flat side rests against the jointer fence. Joint a second side, holding the wood flat against the fence as you cut.

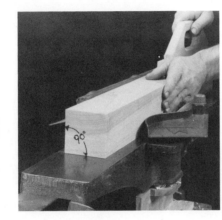

1/You must square the leg stock before you cut the legs. To do this, check that the jointer fence is precisely square to its table. Then joint one face of the board perfectly flat.

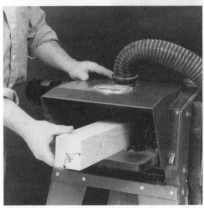

3/Mark the jointed faces of the leg stock with large Xs. Plane the faces of the legs **opposite** the Xs, removing small amounts of stock until the legs are 2¾" square.

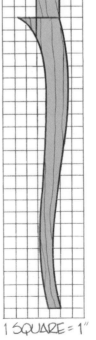

SIDE VIEW

21"
1"
19"
1"
2"

20⅛"

LEG AND EAR PATTERN

1 SQUARE = 1"

Using a square, draw lines on all four faces of each leg precisely 4″ below the top end — this is where the post of each leg should end. Select the two *inside* faces of each leg. Using the template, trace the shape of the leg on both of these faces. (See Figure 4.) Then mark the mortises in the post ends of the legs.

Cut the mortises with a table-mounted router and a ¼″ straight bit, then square the blind ends of the mortises with a chisel. (See Figure 5.) After making the mortises, glue the ear blocks to the legs, as shown in the *Mortise and Ear Block Detail*. After the glue dries, trace the shapes of the ears on the *inside* surfaces of the blocks, as you did for the legs.

MORTISE AND EAR BLOCK DETAIL

4/Trace the shape of the legs on the two **inside** faces of the stock, using the template. To make sure each shape is properly aligned with the other, line up the horizontal marks on wood with the one on the template.

5/Rout the mortise in the legs on a router table, using a straightedge or a fence as a guide. Feed the stock so the rotation of the bit helps hold it against the fence. Make each mortise in several passes, cutting just ⅛″–¼″ deeper with each pass.

3 **Cut the curve of the legs.** Cut the shape of the legs and ears on a band saw, following this procedure:

Turn a leg so one ear block points up. Cut one face of the post, from the top end down to the knee. Adjust the upper blade guide of the band saw to clear the ear, and cut the knee and the outside surface of the ear, freeing the waste from the post. Readjust the blade guide. Starting at the bottom of the stock, cut the rest of the leg shape. Save all the scrap, and tape it back to the stock. Turn the leg 90° so the other ear points up, and repeat this sequence of cuts. When you remove the tape and the waste, you'll have a gracefully curved leg. (See Figures 6 through 10.)

Sand, file, and scrape the surfaces of the legs smooth. Carefully preserve the crisp lines between the surfaces — don't round over the corners.

6/Cut the shape of each leg with a band saw. First, position the leg stock so one ear block points up and the other lies flat on the table. Cut one face of the post, down to where the knee begins.

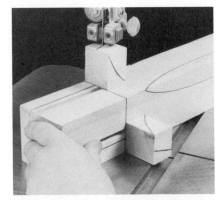

7/Adjust the upper blade guide to clear the ear block. Cut the outside surface of the knee and the ear, freeing the waste from the post. Save the waste.

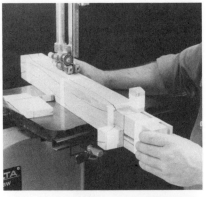

9/Tape the waste back to the leg stock, making the leg rectangular again. Turn the stock so the other ear block points up, and repeat the same sequence of cuts.

8/Readjust the blade guide and cut the remainder of the leg shape, both the inside and outside surfaces. Save all the waste.

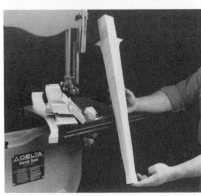

10/When you remove the tape and the waste, you'll have a gracefully curved leg with pointed ears.

4 **Make the aprons.** Cut the front apron ¼″ wider and ½″–1″ longer than specified in the Materials List. Using a saw that cuts a ⅛″-wide kerf, rip ½″-wide strips from the top and bottom edges. The remaining stock should be 3″ wide. From the middle of this piece, cut a board 17⅞″ long. Set this aside for the drawer front.

Glue the ½″-wide and 3″-wide pieces back together, leaving a 3″-wide, 18″-long opening in the center of the apron. Let the glue dry, then cut the apron to length. Make sure you remove the proper amount of stock from each end, so the opening is centered precisely. Rip and cut the back and side aprons to the sizes given in the Materials List.

The tabletop is attached to the aprons with clips, as shown in the *Top-to-Apron Joinery Detail*.

While the blade that cuts a ⅛″-wide kerf is still mounted on the saw, cut a ⅛″-wide, ¼″-deep groove in the *inside* face of each apron, approximately ¼″ from the top edge. The precise distance from the top edge depends on the tabletop clips you're using.

Using a dado cutter, cut ¼″-wide, 1″-long tenons on the ends of all the aprons. (See Figure 11.) These tenons should fit snugly in the leg mortises.

11/Cut the tenons in the ends of the aprons with a dado cutter mounted on a table saw or radial arm saw. Attach a stop block to the saw fence to help position the boards on the saw. Set up the cutter and the stop block to cut a 1″-wide, ¼″-deep rabbet. Cut a rabbet in one face of a board, then turn it over and cut a rabbet in the opposite face. The two matching rabbets will form a tenon.

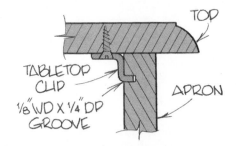

TOP

TABLETOP CLIP

⅛″ WD X ¼″ DP GROOVE

APRON

TOP-TO-APRON JOINERY DETAIL

5

Make the guides and kickers. Cut the drawer guides and kickers to size. Using the dado cutter, make a ½"-wide, ⅝"-deep rabbet in the inside edge of each guide, as shown in the *Drawer Guide Detail*.

Using a doweling jig, drill two ¼"-diameter, ½"-deep dowel holes in each end of each drawer guide and kicker. Measure and mark where these parts will join the front and back aprons. Measure carefully; you must mark these positions as precisely as possible. Insert

dowel centers into the holes in one end of a guide or kicker, then press it against an apron where you want to mount it. (See Figure 12.) Repeat for each end of each piece, then drill matching dowel holes in the aprons.

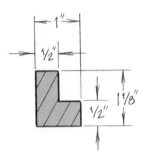

DRAWER GUIDE DETAIL

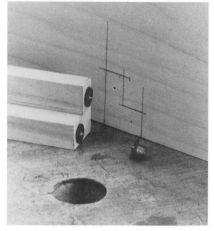

12/Drill the dowel holes in the drawer guides and kickers, then use dowel centers to locate the corresponding holes in the aprons. These centers leave small indentations in the wood, showing you where to drill the holes. Carefully mark all the pieces — indicating which end joins which apron — so you won't get them mixed up.

6

Make the top. Glue ¾"-thick boards edge-to-edge to create wide stock for the tabletop, then cut it to size. Enlarge the *Tabletop Pattern* and trace it

on the stock. Cut it out with a saber saw and sand the sawed edges. Using a router and a ½" quarter-round bit, shape the edge of the table as shown in the *Edge Detail*.

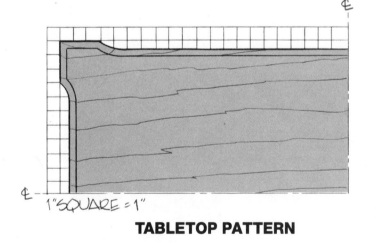

TABLETOP PATTERN

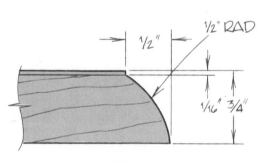

EDGE DETAIL

7

Assemble the table. Finish sand the legs and aprons. Once again, be careful not to round over the edges.

Glue the front and back aprons to the legs. Allow the glue to set, then lay the back apron/leg assembly on the

floor of your shop, with the inside surface facing up. Attach the drawer guides and the kickers with glue and dowels, then glue the side aprons to the back legs. Before the glue dries, attach the front apron/leg assembly.

Stand the table assembly up and clamp it together. As you tighten the clamps, check that the parts are square to each other. In particular, make sure the drawer guides are square to the front apron. If a part needs adjustment, loosen the clamps and shift it before the glue dries.

After the glue sets, put the tabletop in place. Put clips in the grooves, spacing them 12"–15" apart. Fasten the clips to the tabletop with flathead wood screws. The clips will secure the top to the aprons, yet allow it to expand and contract with changes in temperature and humidity.

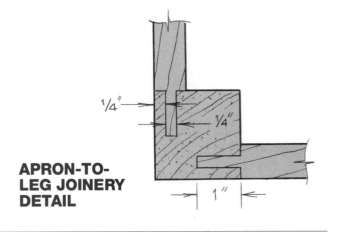

APRON-TO-LEG JOINERY DETAIL

8 ***Make the drawer.*** Cut the parts of the drawer to size, using the piece cut from the front apron as the drawer front. As you build this drawer, remember to orient the front so the grain pattern matches that of the apron. When the drawer is in place, the apron and the drawer front should appear to be one continuous board.

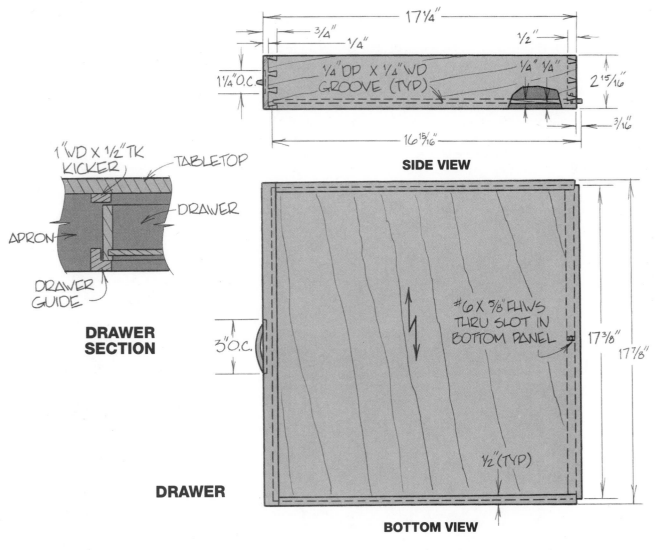

SIDE VIEW

DRAWER SECTION

DRAWER

BOTTOM VIEW

As shown, the parts of the drawer are joined with dovetails. If you cut these by hand with a chisel and a fine saw, make the pins in the front and the back first. Then use the pins to mark the shapes of the tails on the sides. You can also use a router and a dovetail jig to cut these joints. Whatever method you use, remember that the dovetails that join the front to the sides are *half-blind*. You mustn't be able to see them when you shut the drawer.

After making the dovetails, cut a ¼″-wide, ¼″-deep groove in the inside face of the drawer front and each drawer side, near the bottom edge. Use a router or a dado cutter to make these grooves. They will hold the bottom in the assembled drawer.

TRY THIS! Professional cabinetmakers often round the top edge of each drawer side with a radius plane or a router and a quarter-round bit. This is optional, but it helps the drawer slide smoothly and gives it a more finished appearance when open.

Dry assemble the drawer to test the fit of the joints and the fit of the drawer in the table. Adjust the joinery and the size of the drawer as necessary.

9 Make the drawer pull and assemble the drawer.

Make a router template, as shown in the *Drawer Pull Template Layout*, from ½″-thick scrap wood. This template will guide the router as you make the cutout for the drawer pull.

Attach the template to the drawer front with double-sided carpet tape. Center the opening in the template on the outside surface of the front, side-to-side and top-to-bottom. Mount a ¼″ straight bit in the router and attach a ⁷⁄₁₆″ guide bushing to its base. Adjust the router to cut ⅛″ deep, then rout the cutout. The guide bushing will ride along the inside edge of template, forming the oval shape. Remove the template and carve the points on the ends of the oval with a chisel. (See Figures 13 and 14.)

Cut the pull slightly oversize. Saw the raised curve of the pull, as shown in the *Pull Detail/Section A,* with a band saw or scroll saw. Sand the sawed edges smooth,

then thin the middle of the pull with a file and sand-paper, as shown in the *Pull Detail/Top View.* Carefully fit the pull to the cutout by filing points on the ends.

Finish sand the drawer parts. Glue the drawer front, sides, and back together. Slide the drawer bottom in place to keep the assembly square while you clamp it together. After the glue dries, screw the bottom to the drawer back.

Note: If you make the drawer bottom from ¼″-thick solid wood instead of plywood, cut a slot for the screw, as shown in the *Drawer/Bottom View.* This will let the bottom expand and contract.

Glue the pull in the cutout. Let the glue set, then sand the ends of the pull flush with the drawer front. Carefully blend the surfaces, so the pull looks to be continuous with the front.

Slide the drawer into the table. If it binds, sand or file away stock until it slides in and out smoothly.

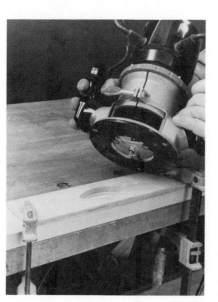

13/To rout the drawer pull cutout, use a template to guide the router. The guide bushing, on the base of the router, rides along the inside edge of the template. This forms a perfect oval-shaped cutout.

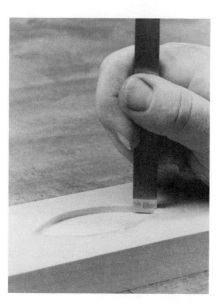

14/Finish the cutout by carving two points at either end of the oval shape.

10

Finish the writing table. Remove the drawer from the table and detach the tabletop from the aprons. Do any necessary touch-up sanding, then apply a finish. Coat all wooden surfaces, except for the inside of the drawer, the kickers, and the drawer guides. (Traditionally, only the fronts of drawers are finished. If you apply finish to other drawer parts, it may interfere with the sliding action.) Be careful to coat the top and bottom of the tabletop evenly — this will prevent it from warping. When the finish dries, wax and buff the surfaces that will show. Then reassemble the table.

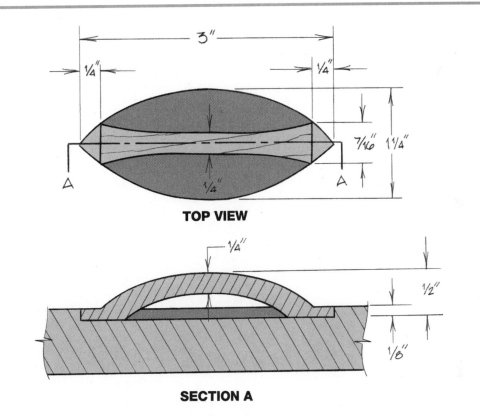

TOP VIEW

SECTION A

PULL DETAIL

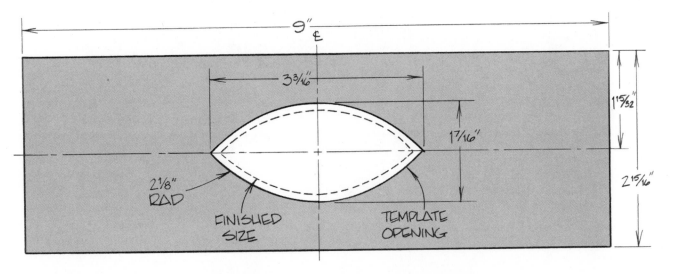

DRAWER PULL TEMPLATE LAYOUT

Hanging Bookcase

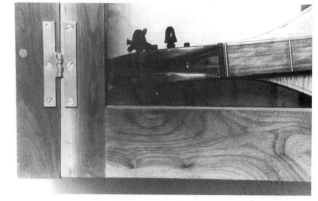

A hanging bookcase offers several advantages over the standing variety. First, you can easily change the height and width to fill the available space. Second, since it's attached to a wall, you can build the case with fewer materials. (The wall serves as the cabinet back and helps keep the case square.) Finally, you can mount the bookcase above a chair or a desk, using otherwise wasted space.

The bookcase shown offers another advantage: It's extremely simple to build. The parts are all rectangular pieces, cut on a table saw. The case is assembled with dadoes and butt joints, and the doors are doweled together. Even the hardware has been simplified — the hinges don't have to be mortised, nor do you need to install door pulls. You don't need to invest much time or materials to build this bookcase, but you can enjoy its utility and contemporary beauty for years.

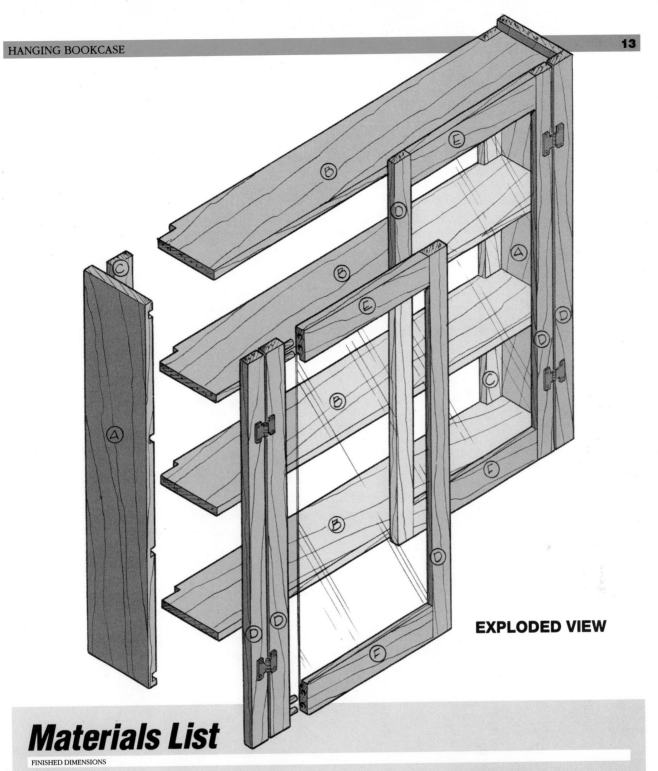

EXPLODED VIEW

Materials List

FINISHED DIMENSIONS

PARTS

A. Sides (2)   ³/₄″ x 8″ x 37″

B. Shelves (4)   ³/₄″ x 8″ x 40″

C. Back stiles (2)   ³/₄″ x 2″ x 35″

D. Front and door stiles (6)   ³/₄″ x 2″ x 37″

E. Upper door rails (2)   ³/₄″ x 2″ x 14¼″

F. Lower door rails (2)   ³/₄″ x 3″ x 14¼″

G. Dowels (16)   ³/₈″ dia. x 2″

HARDWARE

#8 x 1¼″ Flathead wood screws (32)

3″ H-hinges and mounting screws (2 pair)

Cabinet door catches (2)

¼″ Molly anchors and roundhead bolts (6)

⅛″ x 14⁵/₈″ x 32³/₈″ Glass panes (2)

Glazing points (8–12)

1

Adjust the size of the bookcase. Measure the space where you want to hang the bookcase and adjust the design to fit this space. You can change the height, width, and depth of the cabinet simply by changing the length or width of parts in the case.

Note: The bookcase shown is approximately 3½' wide. This is the maximum width for the case *as designed*. If you make it any wider, the shelves may sag in the middle. To make a wider bookcase, you must modify the design beyond simply changing the measurements. Divide the case up into sections and add interior walls or dividers to help support the shelves every 3' to 4'. Also add back stiles wherever you put dividers to help support the extra weight of the larger case.

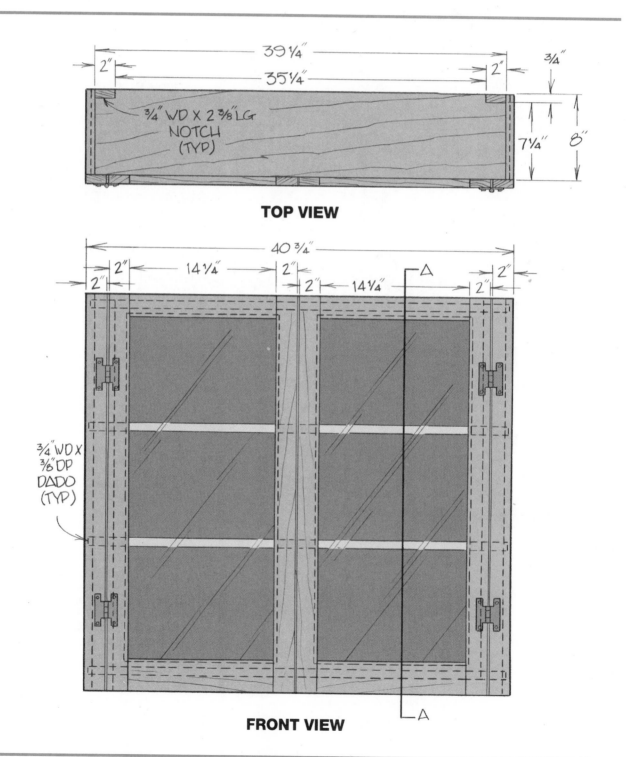

TOP VIEW

FRONT VIEW

2 **Cut the parts to size.** To build this project as drawn, purchase approximately 20 board feet of cabinet-grade lumber. If you change the dimensions, remember that the amount of material will change, too. You can easily calculate the number of board feet you'll need by figuring the number of *square feet* in all the parts and adding 10 percent extra for waste. Because all the parts are ¾″ thick (planed down from 1″-thick or four-quarters stock), the number of square feet will be equal to the number of board feet. (A board foot is 144 cubic inches of lumber.)

Once you have purchased the lumber and planed it to the proper thickness, cut all the parts to size.

3 **Cut the joinery in the sides.** Using a router or a dado cutter, make ¾″-wide, ⅜″-deep dadoes in the sides, as shown in *Section A*. To save time and ensure that you make both sides exactly the same, clamp the sides together edge to edge. Cut the joinery in both at the same time. (See Figure 1.)

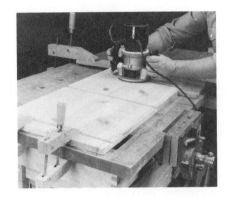

1/When routing the dadoes in the sides, clamp both boards edge to edge on your workbench. Clamp a straight-edge to guide the router across the boards, then rout both the right and the left side at the same time.

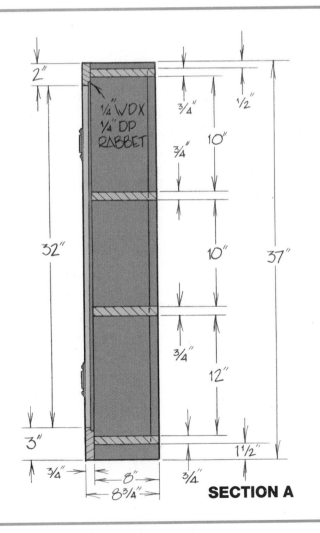

SECTION A

4 **Cut notches in the shelves.** Set the back stiles in notches in the shelves, so they will be flush with the back edge of the case. Cut these 2⅜″-wide, ¾″-deep notches with a hand saw or saber saw, as shown in the *Top View.*

5 **Assemble the case.** Dry assemble the sides, shelves, and front and back stiles to test their fit. Make minor adjustments, if necessary. When you're satisfied that the parts fit properly, disassemble the bookcase. Finish sand all the parts.

Assemble the sides and shelves with glue and screws. Drive the #8 x 1¼″ flathead wood screws at an angle up through the shelves and into the sides, as shown in the *Shelf-to-Side Joinery Detail.* This will hide the screws, making them invisible from the outside of the assembly.

Secure the back stiles in their notches with glue and screws, driving the screws through the stiles and into the shelves. Countersink the heads of the screws so they are flush with the back surface of the stiles. Glue

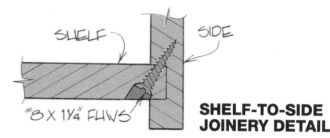

SHELF-TO-SIDE JOINERY DETAIL

and screw the front stiles to the sides and shelves, counterboring *and* countersinking the screw heads. Cover the heads with wooden plugs and sand the plugs flush with the wood surface.

6 **Drill the door for dowels.** To lay out the dowel holes, first dry assemble the door parts on your workbench and clamp them together. Draw lines across the face of each butt joint, marking both the rail and stile wherever you want to put a dowel. Disassemble the frames, then drill ⅜″-diameter, 1⅛″-deep dowel holes at each of the marks with a doweling jig. (See Figure 2.)

Note: The door frame stock must be perfectly straight, with no warps, bends, or twists. *This is extremely important!* If the frames are even slightly distorted, the doors will not fit the case correctly.

2/Use a doweling jig to drill dowel holes in the ends of the rails and edges of the stiles. To position each hole, line the jig up with a mark on the wood. To stop the hole at the proper depth, attach a stop collar to the drill bit.

7 **Assemble the door frames.** Dry assemble the door frames with dowels to check the fit of the joints. Also check the fit of the doors in the bookcase. Each door should be the same height as the case, and when laid side by side, the two doors together should be slightly narrower than the width of the opening. When you're satisfied the doors and the door joinery fit properly, disassemble the frame. Finish sand the parts, being careful not to round over any adjoining edges or ends.

Assemble the frames with glue. Let the glue dry and sand the joints clean and flush. Clamp the frames to the workbench, with the outside surface down. Using a

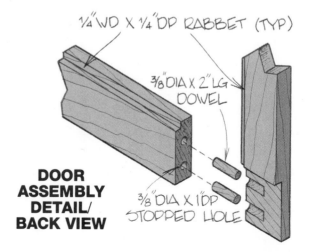

DOOR ASSEMBLY DETAIL/ BACK VIEW

¼″WD X ¼″DP RABBET (TYP)

⅜″DIA X 2″LG DOWEL

⅜″DIA X 1″DP STOPPED HOLE

router, cut a ¼"-wide, ¼"-deep rabbet all around the inside edges of each door frame. (See Figure 3.) This rabbet will hold the glass in the door. Square the corners of the rabbet with a chisel.

3/Rout a rabbet for the glass in the inside edges of the door frame. Before you rout the wood, let the dowel joints dry for at least 24 hours to attain full strength. Keep the frame clamped securely to the bench as you cut.

> **TRY THIS!** Many cabinetmakers prefer to build the doors slightly oversize, then plane them or sand them to the size needed. This enables them to get a perfect fit, even if there are small defects in the door or the case.

8 Mount the doors on the bookcase. Lay the bookcase on its back, then lay the doors in the bookcase. If you have made the doors oversize, plane or sand them to fit the opening. Arrange them so there is a small gap between each of the doors and between the doors and the front stiles. These gaps should be approximately ¹⁄₁₆" wide.

Put the hinges in place, straddling the gap between the doors and the front stiles. Mark the position of the hinges, drill pilot holes for the mounting screws, and attach the hinges. Attach friction or magnetic door catches to the lower door stile and the underside of the bottom shelf. Then check the action of the doors, making sure they open easily and latch securely.

9 Finish the bookcase. Remove the doors from the bookcase and detach all the hardware. Do any touch-up sanding necessary to ready the project for finishing. Apply a finish to the case and door frames. As you work, make certain to cover both the inside *and*

the outside of the case equally. If you apply less finish to one side of a board, it will absorb more moisture than the other side. This may cause the board to warp or cup, ruining the project.

10 Mount the bookcase on the wall. For safety, attach at least one side of the bookcase to the wall frame, using #14 x 2½" roundhead wood screws and flat washers. If the other side rests over a hollow wall between two frame members, use Molly anchors to attach that side. When attaching a bookcase to a masonry wall, use ³⁄₁₆" x 2½" lag screws and expandable lead anchors.

With a helper or two, position the bookcase on the wall. Make sure that one of the back stiles is over a

stud. Make sure *both* stiles are over a stud if the case is particularly large or heavy. Drill one pilot hole through that stile and into the stud. Install a screw in the hole.

Level the case, pivoting it on the screw, then drill the remaining holes needed. Remove the screw and set the case aside. Install Molly anchors or expandable lead anchors in the wall, if you're using them. Replace the case on the wall, and drive screws or bolts through the mounting holes to secure it.

11 Install the doors and the glass. When you've hung the bookcase, replace the doors and reattach the hardware. Mount panes of glass in the

doors, using glazing points to keep them in place. To prevent the glass from rattling in the frames, put a small piece of felt between the glass and each glazing point.

Desk Carousel

A desk is not just a place to work, it's also a place to meditate. Many people find that their best work is preceded by careful thought, and to nourish this reflection they keep a few "executive toys" at their desks. These are often slick, designer versions of childhood toys — yo-yos, marbles, tops, and so on — that move in some - uncomplicated, controlled manner. Their simple motion can have a soothing, hypnotic effect that promotes a contemplative state for problem solving.

This desk carousel is one such toy. It has a practical side, too: The top of the carousel provides a place to keep letters, paper clips, stamps, and other paperwork paraphernalia. However, it is apt to get more use as a miniature merry-go-round.

"It's actually a white-collar prayer wheel," says Judy Ditmer, an award-winning lathe turner who designed and turned the carousel at Heartwoods, her woodworking shop in Tipp City, Ohio. "It's pleasant to spin it around and around as you sit and think, like the Buddhists of Tibet spin their prayer wheels. And when you get tired of the carousel, the lids on the paper clip and stamp containers

become tiny wooden tops — you can spin them!" There are wheels within wheels....

Note: This is a *faceplate turning* project. With just one exception, all of the parts are shaped on a lathe using a faceplate turning accessory.

Materials List

FINISHED DIMENSIONS

PARTS

A.	Base	6¾"-dia. x 1"
B.	Platform	9¼"-dia. x 1"
C.	Large dividers (2)	8"-dia. x 1"
D.	Small divider	5"-dia. x 1"
E.	Paper clip/stamp holders (2)	1⅝"-dia. x 1¾"
F.	Pencil holder	1⅝"-dia. x 3½"
G.	Tops (2)	1¾"-dia. x 1½"
H.	Dowels (9)	¼"-dia. x 1"

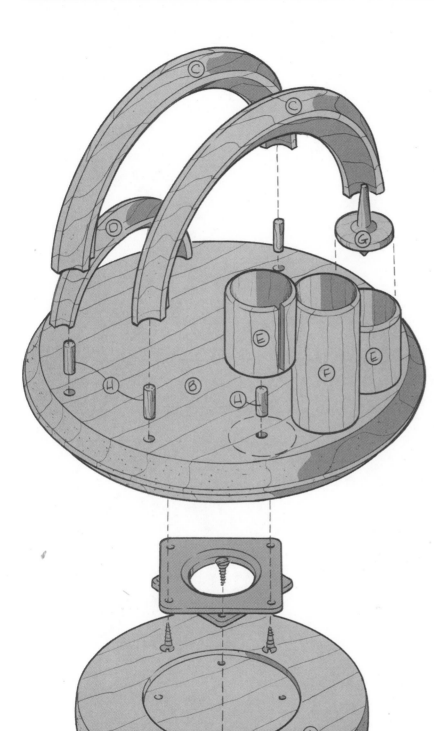

EXPLODED VIEW

HARDWARE

4" Lazy Susan swivel and mounting screws

Felt (6" x 6")

1 ***Select the stock and cut the parts.*** To make this project, you need approximately 3 board feet of 5/4 (five-quarters) stock, and less than 1 board foot of 8/4 (eight-quarters) stock. The carousel shown is made of padauk, but you can use any hardwood. You can also mix hardwoods, if you wish. The parts are small enough that you may be able to build the carousel entirely from scrap wood. This may present an opportunity to use some of those small pieces of wood that were just too pretty to throw away.

Plane the wood about ⅛″ thicker than specified in the Materials List. Cut the parts with a band saw, making round disks and cylinders. Make the disks approximately ½″ larger in diameter than specified, and make the tops 1″–2″ longer. You'll need the extra stock to true up the parts after you mount them on the lathe face-

plate. The top stock must be extra long so you have room to work between the lathe centers. Trim the holders and tops to octagons on the table saw. Cut the dowels to the sizes given.

TRY THIS! When you turn the base, platform, and dividers, the grain will be perpendicular to the axis of rotation. Consequently, the chisel wants to dig into the wood, and the pieces are hard to turn. To help prevent this, use extremely dense hardwoods for these parts, such as padauk, rosewood, teak, cherry, and rock maple.

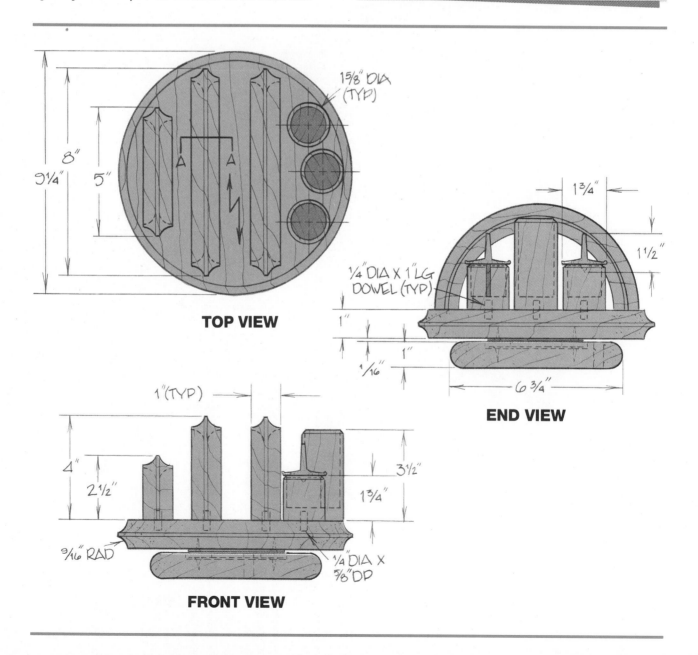

TOP VIEW

END VIEW

FRONT VIEW

2 Turn the base and platform.

Turn the base and platform. Mount the base stock to a faceplate and turn it to the proper diameter. Be sure the screws that attach the stock to the faceplate won't extend through the stock once you've turned it to shape.

Scrape the edge to shape with a skew chisel, as shown in the *Front View*. Finish sand the rounded edge on the lathe. Also, turn a recess in the face of the base. This recess must be wide enough to house the lazy Susan swivel, and *almost* as deep as the swivel is thick. When you assemble the carousel, the swivel should protrude approximately 1/16″ above the top surface of the base. (See Figure 1.)

Turn the platform using the same procedure. Mount the stock to a faceplate, turn it to the proper diameter, and cut a shape in the edge. Instead of rounding the edge, use a gouge or a roundnose chisel to cut two

1/Turn a recess in the top surface of the base to fit the swivel. The bottom of this recess must be as flat as possible.

coves. Create one cove on the top of the base and the other on the bottom, as shown in the *Front View*. Finish sand the edge while the part is still mounted to the faceplate.

3 Make the dividers.

Make the dividers. Mount the divider stock to the faceplate, turn it to the proper diameter, and cut two coves in the edge — as you did when making the platform. Then cut two more coves, just inside the first pair, to form the star shape shown in *Section A*. Do not cut through the disk completely just yet. Finish sand the coves first. Then, turning at a low speed and using a parting tool, carefully cut through the last 1/8″ of stock. Catch the ring as it separates from the waste. (See Figures 2 through 4.) Turn a second, smaller

ring from the leftover stock, cutting precisely the same shape in the ring.

Sand the insides of the rings, then cut them in half on a band saw. You must cut the rings *parallel* to the grain direction, so the grain in each half-ring divider will run horizontal when you glue it to the platform. *This is very important!* If the grain isn't properly oriented, the dividers may break. True the cut ends with a disk sander or stationary belt sander, so the dividers will sit flat on the platform. (See Figures 5 through 7.)

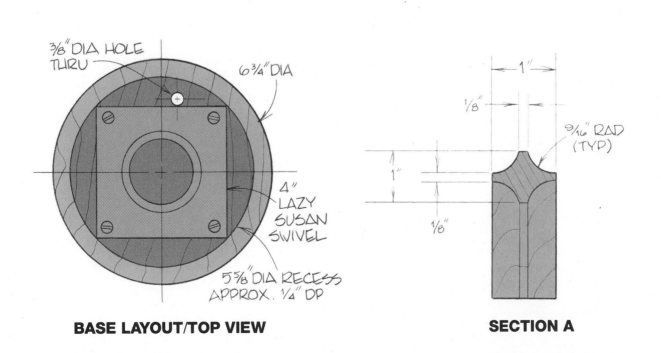

BASE LAYOUT/TOP VIEW

SECTION A

2/Make the dividers by turning rings on the lathe. To make a ring, turn a disk to the proper diameter, then cut the shape of the ring in the disk. Sand the ring, then separate the ring from the disk.

3/Using a parting tool, cut through the last little bit of stock, until the ring is free of the disk. Turn at a very slow speed and feed the chisel with a gentle pressure.

4/When the ring comes free, catch it before it hits the floor. This is why you must turn the last part of the ring at a slow speed: If you turn too fast, the lathe may fling the ring across the room when it separates from the disk.

5/Sand the ⅛"-wide flat inside each ring by hand or use a drum sander. Be very careful not to remove too much stock from any one area. If you do, the inside flat won't be even all the way around the ring.

6/Cut each ring precisely in half on a band saw. Make the cut parallel to the grain direction of the ring.

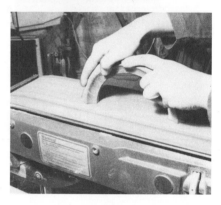

7/Sand the cut ends of each half-ring divider on a stationary disk sander or belt sander, so the dividers will sit perfectly flat on the platform.

4   Turn the holders. If you have a three-jaw chuck for your lathe, mount the holder stock in it. If not, glue and clamp the stock to a small scrap disk. Put a *single sheet* of newspaper between the stock and the scrap as you glue them up — this will help you separate the completed holder from the scrap. Let the glue dry overnight, then mount the scrap to a faceplate.

Turn the inside of the holder first, using a gouge and a flat scraper. Make the sides straight and the bottom flat. Turn the outside with a skew chisel until the walls are just ⅛" wide. Finish sand the holder on the lathe, then separate the holder from the scrap. (See Figure 8.) Make three holders, one tall and two short.

Using a coping saw, cut a ⅛"-wide slot lengthwise in one short holder, as shown in the *End View*. You can dispense rolls of stamps from this slot. Smooth the edges of the slot with a small flat file.

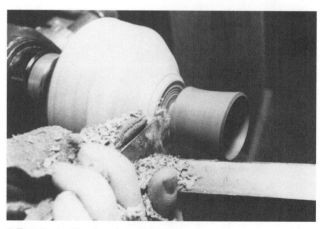

8/Turn the inside of each holder first, then the outside. Measure the thickness of the walls from time to time as you work, so you don't make them too thin.

5

Turn the tops. The tops are the only parts that you don't turn using a faceplate — turn them as *spindles*. To make each top, mount the stock between centers. Using a small gouge and a skew chisel, turn the shape shown in the *Top Detail/Section View*. (See Figure 9.) Finish sand the shape on the lathe, then separate it from the waste with a parting tool.

*9/Make the tops as **spindle** turnings. Use a very gentle pressure to feed the chisels; otherwise, you may snap the long, slender handles on the tops.*

6

Assemble the platform, dividers, and holders. Drill ¼"-diameter, ⅜"-deep holes in the bottoms of the holders and the flat ends of the dividers. Carefully mark the position of the holders and dividers on the platform. Insert a dowel center into the

bottom of each part and press it in place on the platform to leave a small indentation in the wood. At each of these indentations, drill ¼"-diameter, ⅝"-deep holes. (See Figures 10 and 11.)

10/Drill dowel holes in the bottoms of the parts you want to mount on the platform. If you have a horizontal boring setup for your drill press, use it as shown to make the holes in the dividers. Clamp the dividers to the worktable to keep them from moving as you drill.

11/Use dowel centers to locate the holes in the platform. Put dowel centers in the holder and divider holes, then press the parts against the platform. The centers will leave small marks, showing you where to drill the matching dowel holes.

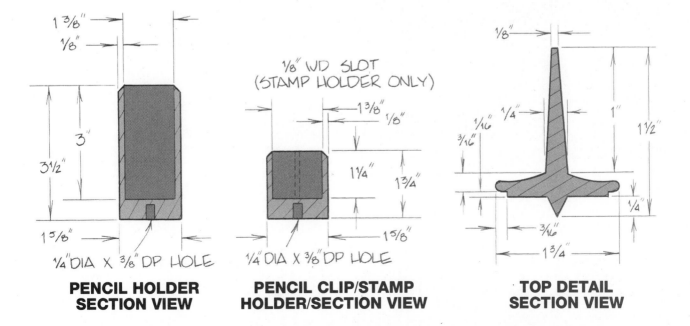

PENCIL HOLDER SECTION VIEW

PENCIL CLIP/STAMP HOLDER/SECTION VIEW

TOP DETAIL SECTION VIEW

Note: When you position the dividers on the platform, make sure the dividers are *parallel* to the grain of the base. The platform will expand and contract *across* the grain with changes in temperature and humidity. If the dividers stretch across the platform perpendicular to the grain, they may break.

Test the fit of the dowel joints by assembling the platform, dividers, and holders without glue. Don't force any parts in place, particularly the dividers. If one of the dowel holes isn't positioned properly, the divider may snap. If a dowel doesn't fit in a hole, enlarge the hole slightly.

When you're satisfied all the parts fit, finish sand the surface of the platform, and do any necessary touch-up sanding on the other parts. Attach the dividers and holders to the platform with epoxy glue.

---

7 *Mount the base on the platform.* Center the swivel on the bottom of the platform and mark the position of all four corner holes. Drill a pilot hole for a mounting screw at each mark.

Place the swivel in the base recess and mark *one* corner hole. Drill a ⅜″-diameter access hole through the base at this mark. Rotate the swivel 45° in the recess, mark the four corner holes, and drill pilot holes. Attach the swivel to the base with screws. The metal swivel must *not* cover the hole you drilled through the base. Place the base on the platform, and line up the swivel with the pilot holes. Reach through the hole in the base with a screwdriver and drive the first screw. Turn the base 90° and drive the second screw. Repeat until you have driven all four screws. (See Figures 12 through 14.)

13/After drilling the access hole in the base, turn the swivel slightly so it doesn't cover the hole. Attach the swivel to the base with screws.

12/Position the swivel on the base, mark one of the corners, and drill a ⅜″-diameter hole through the base at the mark. This will serve as an access hole, allowing you to attach the swivel to both the platform and the base.

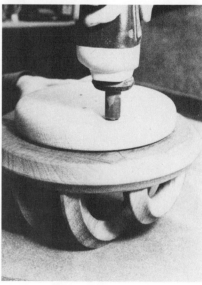

14/Center the base on the platform with the swivel between the two parts. Attach the swivel to the platform, reaching through the access hole in the base to drive the screws.

---

8 *Finish the carousel.* Remove the base from the platform and detach the swivel. Do any necessary touch-up sanding on the pieces, then apply a finish to the platform assembly, base, and tops. When the finish dries, reassemble the swivel, base, and platform. Cut a 6″-diameter circle of felt and attach it to the bottom of the base with contact cement. Turn the carousel right side up and place the tops on the small holders.

Computer Workstation

A computer isn't just a single machine. It's actually a *system* of electronic components — central processing unit, monitor, keyboard, disk drives, and printer. In addition, it requires cables, manuals, printer paper, diskettes, and other peripheral materials. To use the system efficiently, you must arrange the components and materials in a single location, where every item is easy to see and reach. Because of this, a special desk has evolved especially for computers, called a *workstation*.

The workstation shown was designed by Jim McCann, an engineer who designs and tests woodworking tools for Shopsmith. Jim also writes and edits woodworking books and magazine articles. For several years, while he used a personal computer for both engineering and literary projects, Jim collected ideas on workstations — designs, dimensions, and special features. This project is a synthesis of those ideas.

Jim built his workstation to be *adaptable,* first and foremost. It will accommodate almost any personal computer system. It's complete — it holds all the necessary equipment and materials, with room left over for software manuals and reference books. It's compact — the workstation will fit in a small space. This, in turn, saves reaching and stretching while you're using the computer. Finally, it's straightforward to build. Most of the parts are simple rectangles, glued and screwed together. There are no fancy joints to cut, drawers to fit, or doors to hang. Yet it's strong, practical, and attractive. ●

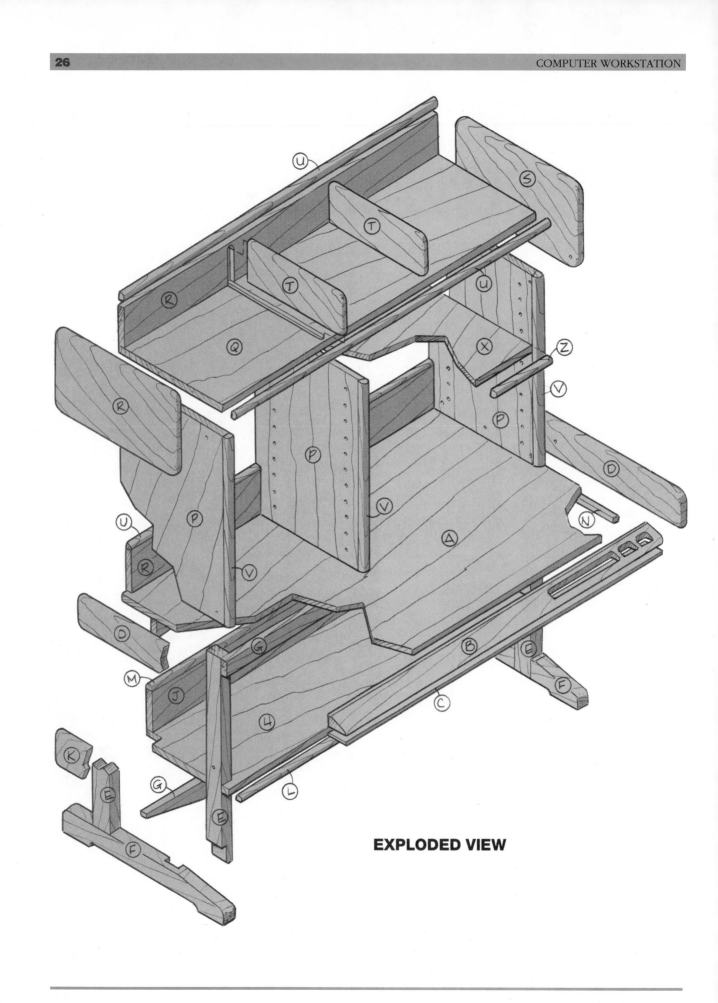

EXPLODED VIEW

Materials List

FINISHED DIMENSIONS

PARTS

Desk Unit

A.	Desk top*	¾" x 28⅞" x 47⅞"
B.	Pencil tray	1¼" x 3" x 47⅞"
C.	Pencil tray bottom	¼" x 3" x 47⅞"
D.	Desk sides (2)	¾" x 4" x 34"
E.	Legs (4)	1½" x 2¾" x 26"
F.	Feet (2)	1½" x 2¾" x 24"
G.	Braces (3)	¾" x 2" x 47⅞"
H.	Lower shelf*	¾" x 13½" x 47⅞"
J.	Lower backstop*	¾" x 4¾" x 44⅞"
K.	Lower shelf sides (2)	¾" x 4" x 16"
L.	Lower shelf edging	¾" x ¾" x 47⅞"
M.	Lower backstop edging	¾" x ¾" x 44⅞"
N.	Cleats (2)	¾" x ¾" x 6"

Shelving Unit

P.	Sides and partition (3)*	¾" x 15¼" x 23⅞"
Q.	Upper shelf*	¾" x 15¼" x 47⅞"
R.	Backstops (2)*	¾" x 6" x 47⅞"
S.	Upper shelf sides (2)	¾" x 10" x 18"
T.	Dividers (2)	¾" x 5" x 14"
U.	Backstop/upper shelf edging (3)	¾" x ¾" x 47⅞"
V.	Side/partition edging (3)	¾" x ¾" x 23⅞"
W.	Short shelves* (1–2)	¾" x 15¼" x 21⅜"
X.	Long shelves* (1–2)	¾" x 15¼" x 24"
Y.	Short shelf edging (1–2)	¾" x ¾" x 21⅜"
Z.	Long shelf edging (1–2)	¾" x ¾" x 24"

HARDWARE

#10 x 1¼" Flathead wood screws (72–84)

#10 x 2½" Flathead wood screws (4–6)

#1 Biscuits or plates (48–60)

Pin-style shelf supports (8–16)

Computer cable grommets (as needed)

Paper grommet (optional)

Multi-outlet surge suppressor and mounting screws (optional)

Make these parts from plywood.

---

1 Adjust the design to fit your system. As shown, the desk will accommodate most standard computer systems — IBM (PC, XT, and AT), IBM clones, Apple, and Commodore. However, before you build this project, measure your system to make sure it will fit. Also, read the section on Computer Ergonomics in this book. Depending on your system and personal requirements, you may wish to adjust the height of the desk top, the placement of the shelves, or other elements of the design.

You should also realize that the desk is not symmetrical; it has a right and a left side. The narrow compartment in the shelving unit holds the computer, keyboard, and monitor. The wide compartment is for the printer, manuals, and other peripherals. Depending on which is your dominant hand, you may want to change the position of these components. The desk in the photo was built for a left-handed person, while the drawings show a right-handed desk. Most right-handed people put their printers on the right; left-handed people do the opposite. The pencil tray will also change sides — it should be opposite the keyboard, so it doesn't interfere with typing.

---

2 Select the stock and cut it to size. To make the computer workstation as shown, you'll need one-and-a-half 4' x 8' sheets of ¾" cabinet-grade (AB) plywood, 20 board feet of 4/4 (four-quarters) solid stock, and 10 board feet of 8/4 (eight-quarters) stock. You may use any combination of plywood and solid wood that is attractive to you. Jim McCann built his workstation from birch-veneer plywood and cherry. These two woods don't ordinarily match, but Jim turned the plywood around so the darker B side shows. When he put a coat of oil on the completed project, the cherry and the dark birch became almost indistinguishable. This trick saved him a great deal of money — cherry-veneer plywood costs over three times as much as birch.

When you have gathered the materials, plane the 8/4 stock to 1½″ thick, and the 4/4 stock to just a hair over ¾″ thick. Since much of the 4/4 stock will be used for edging, you want to make it slightly thicker than the plywood. After you glue it in place, you can scrape and sand it flush with the plywood. If you have to scrape and sand the plywood instead, you may wear through the veneer.

Cut all the parts to the lengths and widths shown in the Materials List, except for the pencil tray and pencil tray bottom. Rip the tray from 1½″-thick stock, and the tray bottom from ¾″-thick stock, making them slightly longer and wider than necessary. Then plane the tray to 1¼″ thick, and the tray bottom to ¼″ thick.

TRY THIS!   To cut large sheets of plywood easily, lay a sheet of fiberboard or subsiding on the floor of your shop and lay the plywood on top of it. Adjust a circular saw to cut through the plywood and just ¹⁄₁₆″ – ⅛″ into the fiberboard. This method is safer and provides more support than laying the plywood across sawhorses.

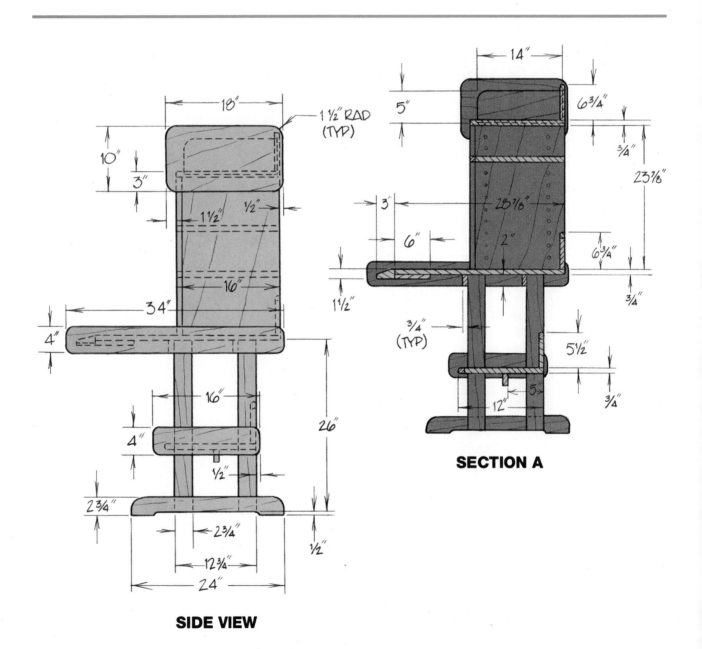

SIDE VIEW

SECTION A

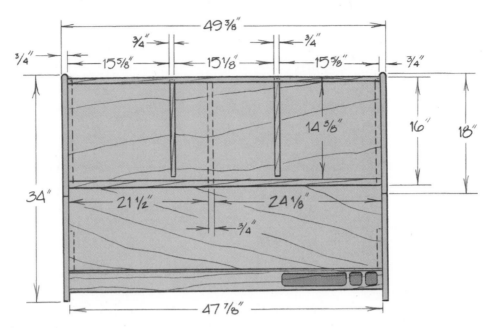

TOP VIEW

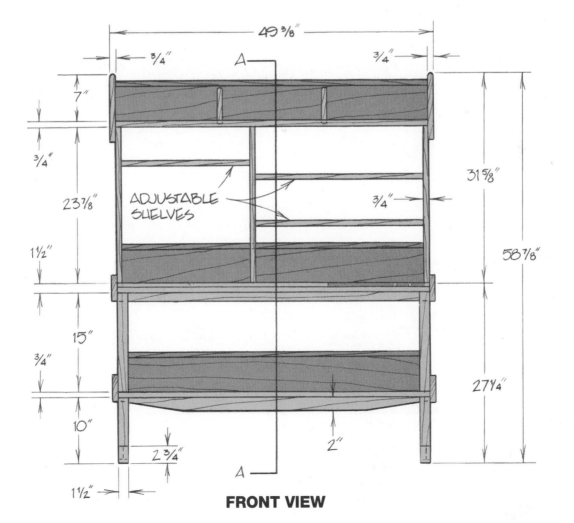

FRONT VIEW

3

Cut the joinery in the legs and feet.
The legs and feet are joined by lap joints, and the lower shelf rests in dadoes in the legs. Carefully lay out the joinery on the stock, as shown in the *Leg Layout* and *Foot Layout*. Cut the lap joints and the dadoes with a dado cutter.

4

Cut the shapes of the feet and the lower brace. Lay out the shapes of the feet and the lower brace, as shown in the *Foot Layout* and the *Lower Brace Layout*. Cut the shapes with a band saw, then sand the sawed edges.

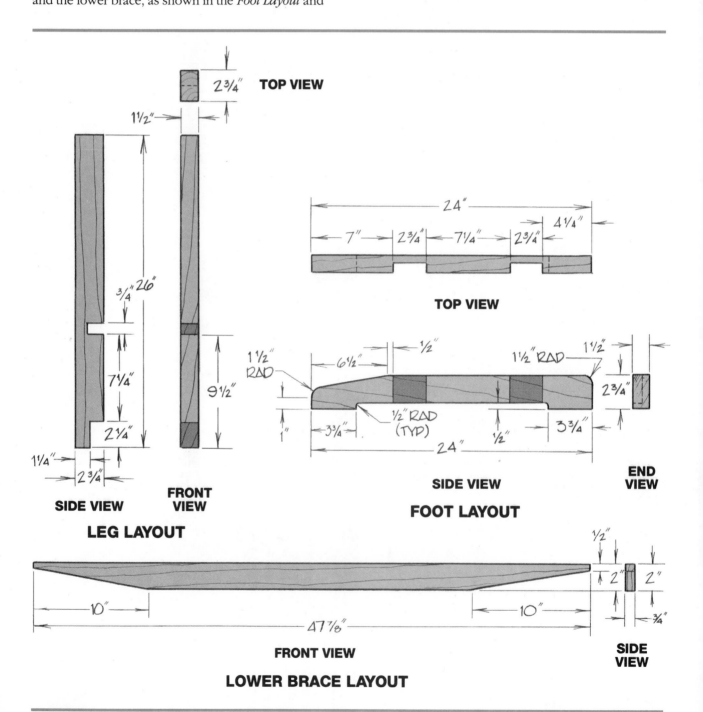

5 Cut and assemble the pencil tray and bottom.

Lay out the compartments in the pencil tray, as shown in the *Pencil Tray Layout/Top View*. To cut out each compartment, first drill a ¾″-diameter hole through the waste. Insert the blade of a saber saw through this hole, and saw to the layout line. Cut along the line, removing the waste. Repeat for each compartment, then sand and file the sawed edges inside the compartments.

Glue the pencil tray bottom to the pencil tray. Before the glue dries, use a wet rag to wipe away any glue that squeezes out inside the compartments. Let the glue set, then cut the pencil tray to the proper length and width.

On a table saw, rip the top surface of the pencil tray to 20°, as shown in the *Pencil Tray Layout/Side View*. Joint the beveled surface to remove the saw marks. (See Figures 1 and 2.)

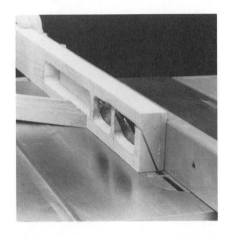

1/The upper surface of the pencil tray is beveled at 20°, so the edge of the desk doesn't bite into your wrists when typing. Cut this bevel on a table saw **after** you've made the pencil tray compartments.

2/Remove the saw marks from the tray by jointing the beveled surface. To help preserve an even, uniform bevel, tilt the jointer fence at 20°.

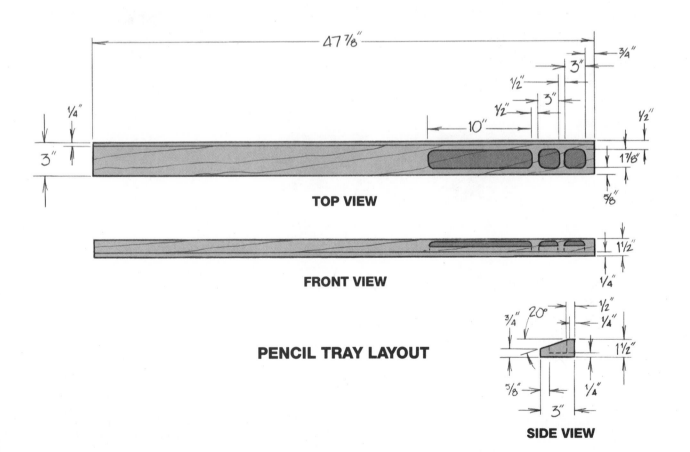

TOP VIEW

FRONT VIEW

PENCIL TRAY LAYOUT

SIDE VIEW

6 Attach the edging to the plywood parts.

Cover the front and top edges of the plywood parts with solid wood so you won't see the plies. Hide the edge of the desk top with the pencil tray, and the edges of the shelves and backstops with 1 x 1 strips. You can attach the tray and strips by several methods — dowels, splines, or tongue-and-groove joints. The easiest method, however, is to use wooden plates or *biscuits*. (See Figures 3 through 6.)

As designed, the plywood edges on the *back* of the computer workstation will show. This shouldn't matter if you place the workstation against the wall so you can't see the back. However, if you intend to arrange it so the back shows, you'll need to glue edging to the fronts *and* backs of the shelves and desk top.

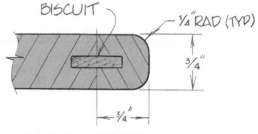

EDGING JOINERY DETAIL

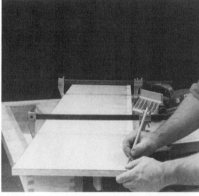

*3/You can join the solid wood edging strips to their plywood parts with plates or **biscuits** in four easy steps. First, put the parts together and draw pencil lines across the joint wherever you want to put a biscuit.*

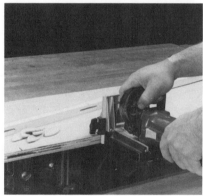

4/Using a biscuit joiner, cut semicircular slots for #20 biscuits at each pencil mark. You must cut matching slots in both the plywood and the solid wood.

5/Apply glue to the adjoining edges and in the slots. Insert #20 biscuits in the slots in the edging, then press the parts together so the other side of the biscuit fits into the slots in the plywood. Clamp the parts together, and wipe away any glue that squeezes out with a wet rag.

6/After the glue dries, scrape the solid wood edging flush with the surface of the plywood. The edging should be slightly thicker than the plywood so there's enough stock to do this. You don't want to scrape the plywood — you may cut through the veneer.

7 Drill holes in the desk top, sides, partition, and backstop, as needed.

The adjustable shelves in the shelving unit rest on movable pin supports. These, in turn, are held in ¼″-diameter holes in the shelving sides and partition. Lay out these holes, as shown in the *Side/Partition Layout*. Remember that you need holes on *both* sides of the partition, but only on the *inside* surface of the sides. Drill the holes in the partition all the way through the plywood, so they show on both sides, but make the holes in the sides just ½″ deep.

In addition to the shelving support holes, you may also wish to make several larger holes or slots in the desk top or backstops for computer cables and paper. The dimensions of these holes will depend on the size of the cable ends, width of the paper, and the measurement of the grommets used to line the holes. Cut the cable holes with a hole saw, and the paper slot with a saber saw.

Note: The grommets are optional, as specified in the Materials List. Their only purpose is to dress up the appearance of the cable holes and paper slot. This becomes more important when you've cut these holes in the plywood parts — you'll probably want to hide the plies. If you cut all your cable and paper holes through the solid wood parts, the grommets are un-

necessary. If you decide not to use grommets, round over the edges of the holes to prevent the cables from fraying and the paper from catching.

8 **Cut the notches in the plywood parts.** Two of the plywood parts require notches. You must notch the lower shelf to fit around the legs. You must also notch the partition to fit around the desktop

backstop. Lay out these notches, as shown in the *Lower Shelf Layout* and *Side/Partition Layout.* Cut the notches with a saber saw.

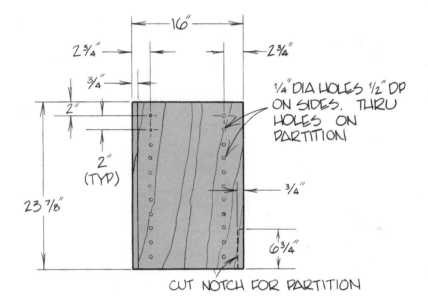

SIDE/PARTITION LAYOUT

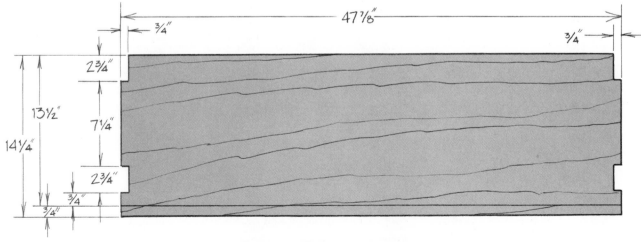

LOWER SHELF LAYOUT

9

Cut the dadoes in the upper shelf and backstop. The dividers rest in dadoes in the upper shelf and backstop. These dadoes are *blind* — closed at one end. Lay out the dadoes on the parts, as shown in the *Upper Backstop Layout* and *Upper Shelf Layout*. Then cut the dadoes with a router and a ¾″ straight bit. (See Figure 7.)

7/To rout a blind dado, first clamp a straightedge to the wood to guide the router. Starting at the back edge, cut toward the front edge, stopping before you exit the wood. Make the dado in several passes, cutting ⅛″ –¼″ deeper with each pass.

TRY THIS! Use a T-square jig to help rout dadoes and rabbets. The long leg of the T is a straightedge, which guides the router. The shorter crosspiece helps to position and align the straightedge.

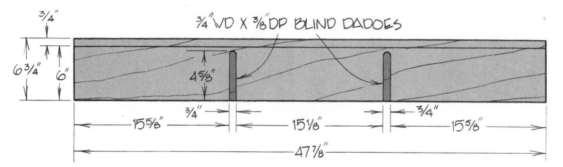

UPPER BACKSTOP LAYOUT

¾″ WD X ⅜″ DP BLIND DADOES

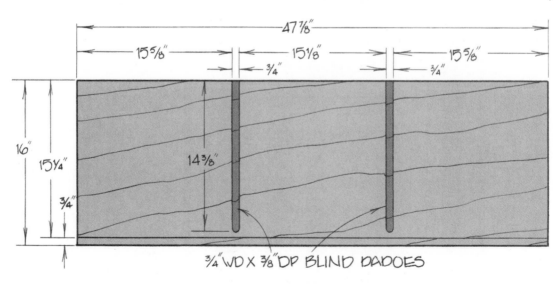

¾″ WD X ⅜″ DP BLIND DADOES

UPPER SHELF LAYOUT

10

Round the edges and corners. Many of the visible edges and corners on this workstation are rounded. This softens their appearance, so the completed piece won't look quite so rectilinear. It also makes the workstation more pleasant to use. Rounded corners are less likely to leave a bruise if you bump into them, and rounded edges are less likely to throw splinters.

To round the corners, cut an 1½"-radius quarter circle with a band saw. To round an edge, cut a radius with a router and a ¼" quarter-round bit. Here is a list of the corners and edges that must be rounded:

Corners
- All four corners of each desk side
- All four corners of each lower shelf side
- All four corners of each upper shelf side
- The upper front corner of each divider

Edges
- All four edges of each desk side
- All four edges of each lower shelf side
- All four edges of each upper shelf side
- The upper edge of each backstop, *except* for the area where the partition joins the desk backstop
- The front edge of each shelf (both fixed and adjustable)
- The front edge of each side and the partition

Sand the sawed and routed edges smooth, then finish sand all the parts you have made.

11

Assemble the desk portion. Assemble the parts of the workstation with glue and screws. Wherever you can, drive the screws from the underside or inside of an assembly so they won't be seen. When you can't hide a screw, counterbore and countersink the screw head, then cover it with a wooden plug. (See Figures 8 through 11.)

To begin assembly, attach the legs to the feet and the lower brace to the lower shelf. Let the glue set, then attach the leg assemblies to the lower shelf. Secure the upper braces to the legs. The top edges of these braces should be flush with the tops of the legs.

Attach the cleats to the underside of the desk top, just behind the pencil tray. The outside edge of each cleat must be flush with the end of the desk top. Secure the desk top to the upper braces, then attach the lower shelf sides and the desktop sides to the assembly.

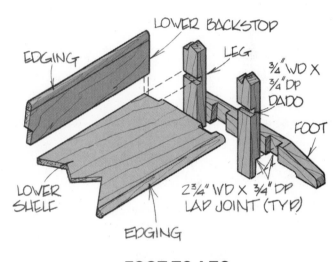

FOOT-TO-LEG JOINERY DETAIL

8/You can make wood screws almost invisible by hiding them under wooden plugs. Use a plug cutter to cut these plugs in scrap wood. This scrap must match the wood in the project as closely as possible.

9/Cut the plugs free from the scrap on a band saw.

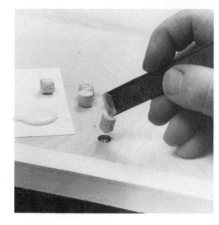

10/Dip each plug in glue, then insert it in a counterbore over a screw head. Be careful to align the wood grain of the plug with that of the surrounding board. Use a chisel to handle the plugs, as shown.

11/Let the glue dry completely, then pare away the top of the plug with a chisel so it's almost flush with the surrounding surface. Sand it down the rest of the way, blending it in with the board. Except for a tiny circle, you shouldn't be able to see any evidence of the plug or the screw.

12 Assemble the shelving unit.

Make the shelving portion of the workstation as a separate unit. Assemble it in the same manner as the desk — with glue and screws — but *don't* glue it to the desk portion. This will allow you to break the workstation down into two parts, should you need to move it or modify it.

Attach the edging to the plywood parts with glue and biscuits. Glue the dividers to the upper shelf, fitting them into the dadoes. Put the upper backstop in place, sliding it over the dividers. Finally, attach the top shelf assembly and the desktop backstop to the sides and partition.

Put the shelving unit in place on the desk, and drive several screws up through the desk top and into the sides and backstop. Do *not* cover the heads of the screws. Leave them accessible, so you can remove them easily.

13 Finish the desk.

Detach the shelving unit from the desk. Soften the edges of the legs and feet with a file and sandpaper so they match the edges elsewhere on the project. They don't need to be as rounded as the routed edges of the shelves, but they shouldn't present hard, crisp lines either. Also, do any necessary touch-up sanding.

Apply a durable finish, such as spar varnish or polyurethane, to the desk, shelving unit, and adjustable shelves. These finishes help protect the wood from metal and plastic computer components. They also resist the wear and tear caused by constantly sliding the keyboard around on the desk top or retrieving manuals from the shelves. Avoid oils and other finishes that penetrate the wood but don't build up a protective coat on the surface.

When the finish dries, wax and buff the desk and shelving units, then put them back together.

14 Install the computer in the desk.

Put the computer workstation in your study or office, several feet from the wall. Put pin supports in the holes in the sides and partition wherever you want them, and rest the adjustable shelves on them.

Put the computer hardware in place on the desk top and the shelves, then run wires and cables between them. If you wish, attach a multi-outlet surge suppressor to the underside of the desk top, just behind the pencil tray. Plug all the components into the suppressor, then plug the suppressor into a wall outlet. This enables you to turn the entire system on or off with one switch, conveniently located just under the desk top. It also protects the delicate electronics from power surges and lightning strikes.

When all the cables are in place, install the cable grommets. If you've cut a paper slot, install the paper grommet. Place a stack of paper beneath the slot, and run the paper up through the slot to the printer. When you have installed the system and it's working properly, slide the workstation against the wall.

Published with the permission of Shopsmith, Inc. Step-by-step instructions on videotape are available from Shopsmith, 3931 Image Drive, Dayton, OH 45414.

Computer Ergonomics

Computers have special furniture needs, as do the people who use them. You can't just plop a computer on a desk, sit an operator in front of it, and expect them to coexist comfortably. The arrangement of the electronic components on the desk is as important as the furniture itself. In this respect, a computer desk is more like an airplane cockpit than a desk. The spatial relationship between humans and their machines has become so important that we have coined a new word for it — *ergonomics*.

There are no hard-and-fast rules in computer ergonomics. The proper spatial relationships change from person to person, and machine to machine. Good ergonomics boil down to this: Do whatever makes you comfortable. When designing your own desk (or *workstation,* as it's usually called), start with a comfortable height for the work surface. For most people, this will be 27″–30″, or about the height of a suitable desk or writing table. Plan to round or bevel the front edge of this surface to make it a more pleasant support for your hands. Then consider where to place each of the computer components, in relation to each other and to you:

CPU — The placement of the CPU (central processing unit) is perhaps the least important of your ergonomic concerns. As long as you can reach the on/off switch, the reset button, and disk drives (if these are included in the CPU box), you can put the CPU almost anywhere. Your only constraint is that it must be central to all the other components, so the connecting wires will reach.

Most operators place the CPU on the work surface, but this occupies a lot of work space that might be better used. Some stand it on its side in a floor stand, next to the workstation. This frees up the work surface, but it may also place the CPU where it could be bumped or knocked over. Also, some disk drives will not work properly when turned sideways — check your owner's manual.

Keyboard — The closer you put the keyboard to your lap, the more comfortable typing is apt to be. If it is placed too high, you have to hold your arms up as you work. This creates fatigue.

Mount the keyboard at the same height as a typing table, approximately 24″–28″. If the work surface is too high to hold the keyboard comfortably, attach a slide-out shelf under the front edge to hold the keyboard. This shelf should lock in place when it's pulled out, so the keyboard won't slide around as you type.

The knee and toe space under the keyboard is also important. Depending on how tall you are, the knee space — the space immediately under the keyboard — should be 20″–26″ high and 12″–15″ deep. The toe space — also under the keyboard, but nearer to the floor — must be deeper, roughly 18″–24″.

Monitor — There are two schools of thought on where to place the monitor. The first (and most common) is to place it at eye level. This position relieves strain on the neck, but it also separates the monitor and the keyboard by as much as 60°, as measured from the eye of the operator. Eye fatigue may result, as you glance back and forth between the two components. If you wear bifocals, or have trouble focusing, this could be uncomfortable.

The second school says to put the monitor as close to the keyboard as possible, the reasoning being that a little neck strain is better than lots of eye strain. Some proponents even suggest building a monitor shelf behind the keyboard surface that is 2″–4″ *lower* than the keyboard. This decreases the angle of separation to 10°–15°, only a little more than the angle between a typewriter keyboard and its platen.

Wherever you place the monitor, the screen should be at least 28″ from your eyes. Like televisions, monitors generate electromagnetic radiation. Although the medical evidence is not yet conclusive, this radiation may cause cataracts and other vision impairments at close range. It may also be harmful to fetuses and pregnant women. To reduce these dangers, position the monitor as far away from you as practical. If you find it difficult to focus on a screen that's 28″ away, consider buying eyeglasses.

Printer — The printer is best placed 18″–24″ above the floor, lower than other components in the system. This lets you reach the printer controls easily *and* puts the platen where you can read the copy as it's printed.

Since the printer is not used all the time, you can store it in a larger drawer or on a slide-out shelf underneath the work surface. Place the paper beneath the printer, in the bottom of the drawer or on another shelf. To operate the printer, slide it and the paper out of the workstation. This saves space, yet still allows you to mount the printer at a comfortable viewing height. You must be careful, however, to turn the printer off before closing the drawer. Otherwise, heat may build up in the drawer and damage the printer.

If the printer is extremely noisy or generates a lot of vibration, you may not want to place it on or in the workstation at all. The noise can be annoying, particularly when printing out a long report or manuscript. And vibration traveling along a wooden workstation may cause other, more sensitive electronic components to malfunction. Disk drives, for instance, are sensitive to vibration. To help isolate the noise and the vibration, build a separate printer stand.

Mouse — If your computer has a mouse, include a large flat space for it next to the keyboard, at the

(Continued)

Computer Ergonomics — Continued

same level. This space should be four to eight times larger than the mouse itself, depending on how sensitive you've adjusted the device to be. The more sensitive the mouse, the less space it requires. If both right- and left-handed people will be using the computer, build space on *both* sides of the keyboard.

Other components — The placement of other computer components and materials, such as disk drives, manuals, and diskettes, isn't as critical. Frequently used items should be within easy reach, either on shelves or in drawers built into the workstation. Arrange things where you won't be tempted to pile other things on top of them. It's all too easy to bury small (but important!) items during an absorbing computing session.

Store seldom-used materials in out-of-the-way parts of the workstation. This will help to make use of otherwise wasted space. For example, operators often build shelves under the workstation, toward the back, to hold old manuals and full data diskettes.

Perhaps the most important ergonomic feature you

can design into any computer workstation is *flexibility*. As much as possible, make it simple to rearrange or adjust the position of the computer components. There are many ways you can do this. Make the parts of the workstation modular, so they come apart and go back together like building blocks. Make the shelves moveable, so you can raise or lower them as needed. Incorporate slides, swivels, and other mechanisms for easy adjustment. Finally, sit in an adjustable swivel chair while working at the station.

There are two good reasons for building in flexibility. The first is comfort: People tend to sit longer at workstations than at desks or on other furniture. You will feel less strain if you can adjust the position of your chair, monitor, or keyboard from time to time.

The second reason is keeping up with rapidly evolving computer technology. You may want to periodically update your system, adding or replacing components. To keep the workstation from becoming obsolete, design it to grow with the system.

Shown here are general measurements for a computer workstation, meant to fit most men and women. However, your own size, working habits, and preferences will determine the best ergonomic arrangement for you. If you're unsure of where to place a computer component, experiment with various positions until one feels right. As much as possible, design the workstation to be flexible and adjustable. This will allow you to rearrange the system as you add new components or your preferences change.

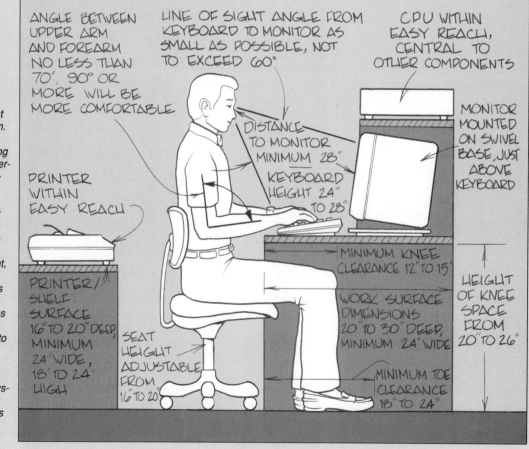

Knife Box Diskette File

Looking for a practical, unique way to protect and organize your computer diskettes? File them in a classic country *knife box!* During the eighteenth and early nineteenth centuries, country folks kept eating utensils in these small wooden containers. There were many types and shapes of knife boxes, but the most popular was a long, rectangular case with a sliding lid, similar to the box shown.

Although most people no longer keep their everyday silverware in boxes, this old design can be modified to store many modern items, including computer diskettes. The dimensions of the box can be easily changed to hold 3½″, 5¼″, or even 8″ diskettes. Wooden dividers help to organize the diskettes, by keeping data apart from programs. You can use any extra space behind the last divider to store diskette labels and other small computer accessories.

Materials List

FINISHED DIMENSIONS

PARTS

A. Sides (2) ½" x 6½" x 12"
B. Tall end ½" x 6½" x 6¾"
C. Short end ½" x 6" x 6¾"
D. Bottom ¼" x 6³⁄₁₆" x 11⁷⁄₁₆"
E. Lid ½" x 6³⁄₁₆" x 11¾"
F. Dividers
(3 or more) ⅛" x 5⅜" x 5½"

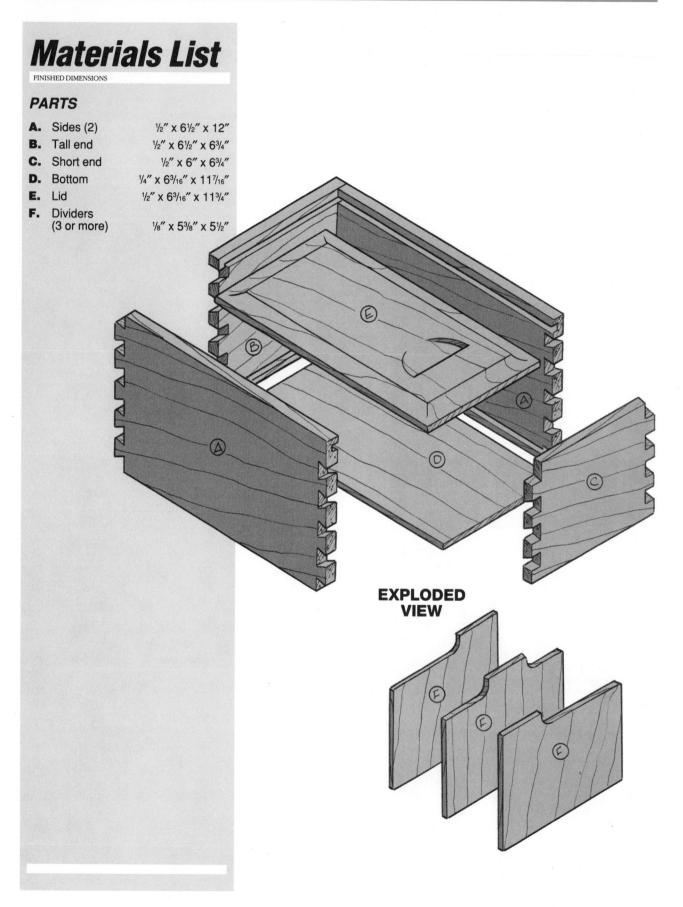

**EXPLODED
VIEW**

1

Select the stock. You'll need three different thicknesses of wood to make this diskette file — ½″ for the sides, ends, and lid, ¼″ for the bottom, and ⅛″ for the dividers. Since all of these pieces are fairly small, you can use up some of the odd scraps that have been sitting around your shop since the last Ice Age. The top of the file shown is made from a highly figured scrap of walnut burl.

You can also mix several species of wood. The different colors and textures emphasize the joinery. The box file shown is made from four woods. The lid and dividers are walnut, the sides are curly maple, the ends are curly birch, and the bottom is birch plywood. When you decide what stock you want to use, plane it to the proper thicknesses and cut it to the sizes shown in the Materials List.

Note: As drawn, the file will hold 5¼″ computer diskettes. If you use 3½″ or 8″ diskettes, change the dimensions accordingly.

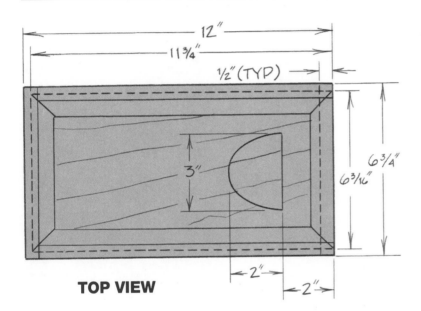

TOP VIEW

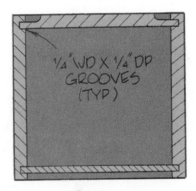

SECTION A

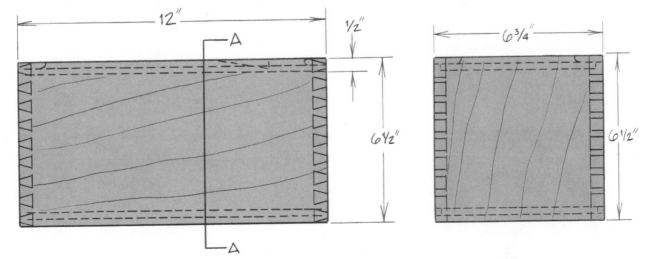

SIDE VIEW

FRONT VIEW

2 Cut the blind grooves.

The bottom and the top are held by blind grooves. All the bottom grooves are double-blind — closed at both ends, as shown in the *Side Layout* and *End Layout*. This prevents you from seeing the grooves when the file is assembled.

The grooves that hold the lid are cut in two different ways. The lid groove in the tall end is double-blind, but those in the sides are open on one end. There is no lid groove in the short end — the top edge is cut even with the bottom face of the lid. This enables the lid to slide open and closed.

Cut all the grooves *before* you make the dovetails in the parts. If you have an overarm router (or an overhead routing jig), you'll find this is the best tool for the job. Mount a ¼" straight bit in the router and clamp a straightedge to the table to guide the workpieces. Carefully mark the blind ends of each groove on the stock. Turn on the router, lower the bit into wood at one end of the groove, and cut until you reach the opposite end. (See Figure 1.) To rout a groove with just one blind end, start the cut at the open end and keep cutting until you reach the blind end.

If you don't have an overarm router, you can use a drill press provided you take certain precautions. An ordinary drill chuck won't support the shank of a router bit against lateral (sideways) thrust. The bit may bend or break. To prevent this, make a collet from a ¼" I.D. bushing by cutting a slot lengthwise along one side. (See Figure 2.) Insert the shank of the router bit in the collet, and clamp the collet in the drill chuck.

Adjust the drill press to run at its highest speed. Use a straightedge to guide and control the cuts, as you would on an overarm router. (See Figure 3.) Make each groove in several passes, routing just ⅟₁₆" deeper with each pass.

When you finish routing the groove, square the ends with a small chisel.

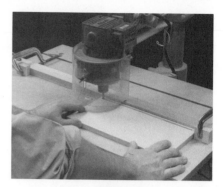

1/If you can, use an overarm router or overhead routing jig to cut the blind grooves. This arrangement gives you more control when routing small parts.

2/To use a drill press as an overarm router, make a special collet to hold the bits. Cut a slot lengthwise along a ¼" I.D. bushing, as shown. Place the bit in the collet, then mount the collet in the chuck. You can find a business that sells bushings by looking in the yellow pages under "Bearings" or "Bushings."

3/When using a drill press to rout the grooves, cut each groove in several passes. The bit may begin to vibrate and chatter if you try to remove too much stock at one time.

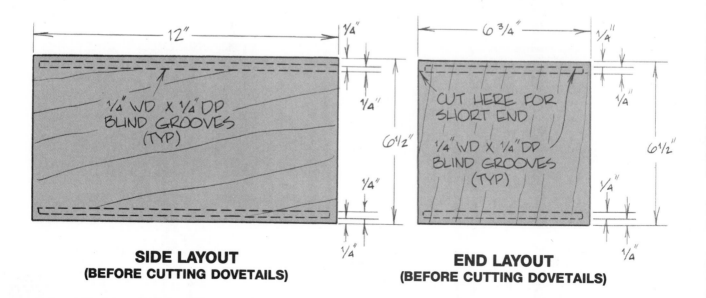

SIDE LAYOUT
(BEFORE CUTTING DOVETAILS)

¼" WD X ¼" DP BLIND GROOVES (TYP)

END LAYOUT
(BEFORE CUTTING DOVETAILS)

CUT HERE FOR SHORT END

¼" WD X ¼" DP BLIND GROOVES (TYP)

3 Cut the dovetails.

Cut the dovetails. Carefully lay out the dovetails on the sides and ends. If one or more of the dovetails are positioned so it will cut through a groove and open a blind end, let it. This will make no difference after final assembly; the newly-opened ends of the grooves will be hidden by the adjoining boards.

The dovetails, however, will show — on both the sides and the ends. You can make these joints, called *through dovetails,* by hand or with power tools. If you make them by hand, cut the pins first, using a fine saw and a chisel. Use the pins as a template to mark the tails, then cut the tails with the same tools.

If you make the dovetail joints with power tools, as shown, use a router and a dovetail jig. Two companies offer special jigs that will make through dovetails. These are available from most mail-order woodworking suppliers, or you can write:

Leigh Industries
P.O. Box 357
Port Coquitlam, BC
Canada V3C 4K6

Keller & Co.
1327 I Street
Petaluma, CA 94952

Both jigs work in a similar manner. A guide bushing mounts on the base of the router and rides against the fingers of the jig. This controls the cut. You cut the tails using a dovetail router bit, and the pins using a straight bit. (See Figure 4.)

Most woodworkers prefer to make the parts of the joint in exactly the opposite order that you would if you were doing them by hand — tails first, then pins. (It's much easier to fit the pins to the tails than it is the other way around.) As you rout, use a scrap block to back up the workpiece and to keep the wood from chipping and tearing. (See Figures 5 and 6.) If the joint doesn't fit correctly, adjust the jig as necessary. Dry assemble the routed parts to check the fit of the dovetail joints. If you've set up the router and the jig properly, you shouldn't have to do any handwork to adjust the fit. (See Figure 7.)

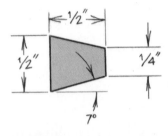

DOVETAIL DETAIL

4/There are two router jigs that enable you to cut through dovetails. In addition to the jig and a router, you also need a guide bushing, a dovetail bit, and a straight bit.

6/Change to the straight bit and the pin template. Rout the pins in the same manner that you cut the tails.

5/Cut the tails first, using the dovetail bit and the tail template. The guide bushing rides against the fingers of the template, guiding the router.

7/Fit the pins and tails together. If the fit is too tight or too loose, adjust the position of the pin template on the stock.

4 **Cut the raised panel in the lid.** The lid is actually a small raised panel. You can raise this panel with the same setup you used to cut the grooves — an overarm router or a drill press. You can also use a table-mounted router.

Using a straight bit, rout a ½"-wide, ¼"-deep rabbet around the perimeter of the lid stock. Switch to a core box bit and cut a small cove in the shoulder of the rabbet. Sand away any mill marks, but be careful not to sand the rabbeted edge much thinner than ¼". If this portion of the lid becomes too thin, it will fit sloppily in the grooves.

Note: If you wish, you can substitute a panel-raising bit for the core-box bit.

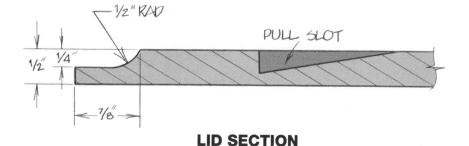

LID SECTION

5 **Carve the pull slot in the lid.** The pull on the lid is a ⅜"-deep tapered slot with one straight side. This slot gives you a purchase for your fingers and allows you to slide the lid open easily.

Lay out the slot on the lid, as shown in the *Top View*. Carve the slot with a chisel and a gouge. To do this, use a chisel to cut ¹⁄₃₂" to ¹⁄₁₆" down *vertically* through the wood, along the straight side of the slot. (See Figure 8.) Switch to the gouge and cut *horizontally,* toward the straight side, removing some of the waste. (See Figure 9.)

Repeat until you have completed the pull slot. Sand the interior of the slot to smooth the gouge marks.

TRY THIS! If you don't want to carve a pull slot, you can rout one. Using a core box bit, cut a 4"-long, ¾"-wide slot across the width of the top, 2" from one end.

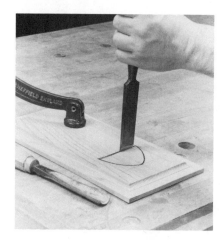

8/To carve the pull slot, first cut vertically along the straight side with a chisel. The chisel must be extremely sharp so it will cut through the wood grain cleanly.

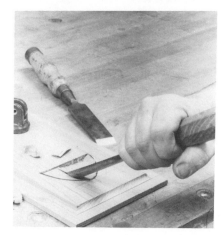

9/Next, cut horizontally with a gouge, pushing the tool toward the straight side. Be careful not to push too hard; you don't want to cut past the straight portion.

6 **Cut the dividers.** Mark the shapes of the wooden dividers on ⅛″-thick stock. Be careful to orient the grain *vertically* through each divider. Cut the shapes on a band saw and sand the sawed edges.

TRY THIS! Like the diskette file itself, the wooden dividers can be made from several different species of wood. The various colors will help you organize and locate the diskettes.

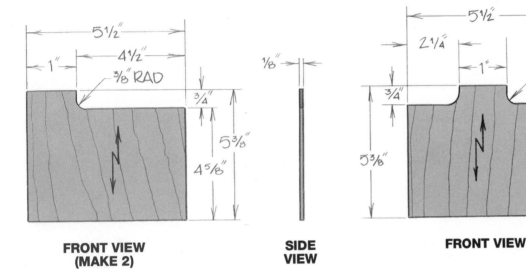

**FRONT VIEW
(MAKE 2)**   **SIDE
VIEW**   **FRONT VIEW**   **SIDE
VIEW**

DISKETTE DIVIDER DETAIL

7 **Assemble the file.** Finish sand all the parts — sides, ends, bottom, lid, and dividers. Test fit the sides, ends, and bottom. The parts should fit snugly, but not too tightly. Also, test the fit of the lid, making sure that it slides easily in the grooves.

When you're satisfied with the fit of the parts, assemble the sides and ends with glue. Put the bottom in place, but don't glue it to the other parts. It should float in the grooves, free to expand and contract. Before the glue dries, slide the lid in place — this will help keep the file square while the glue sets. However, be careful not to get any glue on the lid or in the grooves. The lid may stick.

8 **Finish the diskette file.** Allow the glue to dry overnight, then sand the dovetail joints clean and flush. Do any necessary touch-up sanding, and apply a finish to all surfaces. Be sure to apply as many coats to the inside of the project as to the outside.

Don't use finishes such as varnish or polyurethane. These may accumulate in the grooves and cause the lid to bind. Instead, apply a *penetrating* finish such as Danish oil or tung oil. Penetrating finishes soak into the wood and won't build up on the surface.

After the finish dries, wax and buff the file. Allow the project to sit for several days with the lid off before you store diskettes in it.

Multipurpose Taboret

Some activities require more tools and materials than you have room for on your desk or worktable. For example, the tools for drafting a woodworking project — rulers, triangles, French curves, compasses, pencils, erasers — take up an enormous amount of desk space. You don't use all of these tools all the time, of course, but you like to keep them handy.

For activities that require lots of small- and medium-size items, you need a taboret. This roll-around cart holds whatever you want wherever you want it. You can place the cart on your right, your left, or behind you, and you can roll it out of the way when you no longer need it. As designed, this taboret will hold many different types and shapes of objects — drafting materials, sewing notions, carving tools, and so on. You can also use it as a typing table, printer stand, microwave cart, even a potting table. You'll quickly find this project is as versatile as it is simple to make. ❋

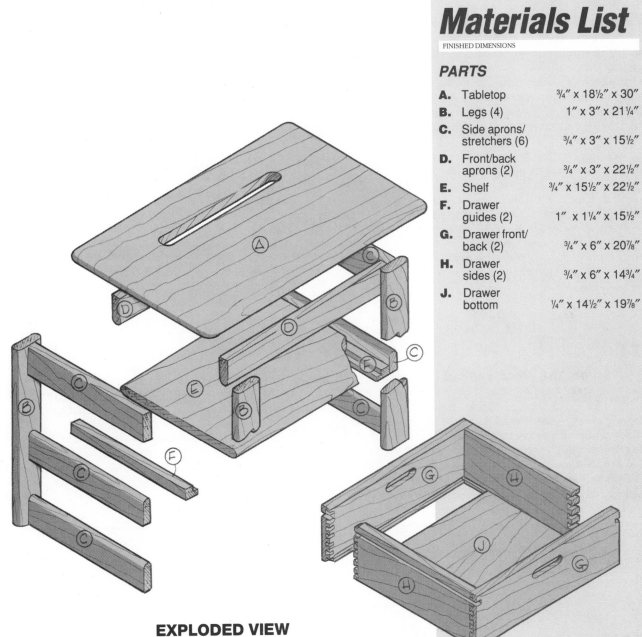

EXPLODED VIEW

Materials List

FINISHED DIMENSIONS

PARTS

A. Tabletop   3/4" x 18½" x 30"

B. Legs (4)   1" x 3" x 21¼"

C. Side aprons/
stretchers (6)   3/4" x 3" x 15½"

D. Front/back
aprons (2)   3/4" x 3" x 22½"

E. Shelf   3/4" x 15½" x 22½"

F. Drawer
guides (2)   1" x 1¼" x 15½"

G. Drawer front/
back (2)   3/4" x 6" x 20⅞"

H. Drawer
sides (2)   3/4" x 6" x 14¾"

J. Drawer
bottom   ¼" x 14½" x 19⅞"

HARDWARE

#10 x 1¼" Flathead wood screws
(30–36)
Tabletop clips (6–8)
#6 x ⅝" Flathead wood screws (6–8)
2" Carpet casters (4)

1 Adjust the design to suit your needs.

As designed, this taboret has three storage areas —
two open areas and one closed. The top holds frequently used tools and materials, while the shelf below
is for seldom-used items. A drawer under the shelf
keeps small, loose objects.

Before you build, give some careful thought to how
you will use this project. You can easily modify the
taboret and any of its storage areas to fit your needs.
For example, if you want to use the taboret as a printer
stand, keep the paper immediately below the printer
and feed it through a slot in the top. (See Figure 1.) If
you want the project to hold sewing notions, build
trays, compartments, and spool racks into the drawer to
help organize needles, thimbles, thread, and other tiny
sewing items. To make a typing table, you will want to
create some leg room under the top. Eliminate the shelf
and the drawer, and add one or two stretchers between
the back legs.

1/This taboret will
also serve as a
printer stand.
Simply cut a slot
in the top to feed
the paper up to
the printer.

2 Select the stock and cut it to size.

To make this project as shown, you'll need one-
quarter of a 4' x 8' sheet of cabinet-grade ¼" plywood
for the drawer bottom. In addition, the legs and drawer
guides require approximately 5 board feet of 5/4 (five-
quarters) stock. The remaining parts take 12 board feet
of 4/4 (four-quarters) stock. Use a hard, dense wood
(such as maple or birch) for the drawer guides. This
stock wears well and will stand up to constant use. The
type of wood used to make the other parts is not critical. Most of the parts of this taboret were made from
white pine, but you can use almost any cabinet-grade
hardwood or softwood that suits you.

After selecting and purchasing the stock, plane the
5/4 wood to 1" thick and the 4/4 wood to ¾" thick.
Glue boards edge-to-edge to make wide stock for the
tabletop and shelf. Then cut all the parts to the sizes
shown in the Materials List.

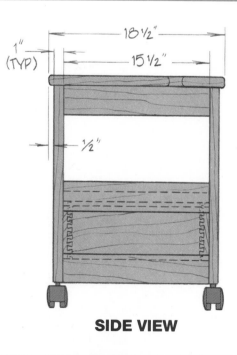

SIDE VIEW

3

Round the edges of the legs and stretchers. Using a router and a ½″ quarter-round bit, round over all four edges of the tabletop and the two vertical edges of each leg. Also, round over the top edges of the middle and lower stretchers. (See Figure 2.) Sand the rounded edges to remove any mill marks. These rounded edges will eliminate some of the hard, straight lines in the design and will soften the appearance of the completed taboret.

2/Round the edges of the top, legs, and stretchers on a table-mounted router, using a ½″ quarter-round bit. The pilot on the bit guides the stock.

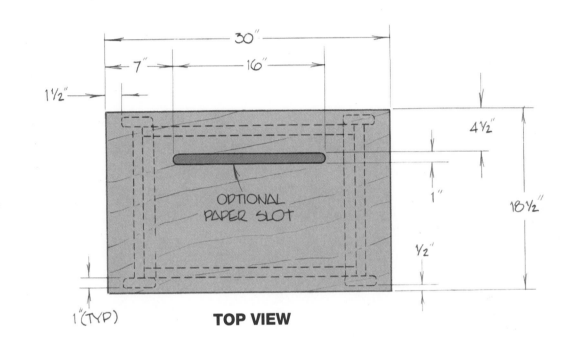

OPTIONAL PAPER SLOT

1″(TYP)

TOP VIEW

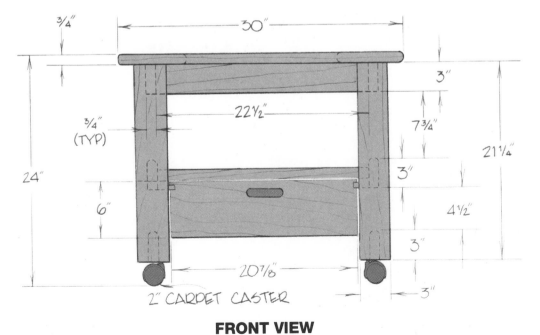

2″ CARPET CASTER

FRONT VIEW

4 Assemble the drawer.

Assemble the drawer. As shown in the drawings, the front, back, and sides of the drawer are joined with half-blind dovetails. You can cut these joints with a router and a dovetail jig. If you don't have a jig, however, you can make the dovetails by hand or substitute lock joints, as shown in the *Alternate Drawer Joinery Detail*.

After cutting these joints, make a ¼″-wide, ¼″-deep groove in the inside face of the drawer front, back, and sides. This groove will hold the bottom, as shown in the *Drawer Detail*.

Dry assemble the drawer to test the fit of the joints. When you're satisfied that they fit properly, disassemble the parts. Cut a slot in the front and the back to serve as handholds. These two slots let you open the drawer from either side of the taboret. To make a slot, first lay it out on the wood. Drill 1″-diameter holes where you've marked the ends, then cut out the waste between the holes with a saber saw. Sand the sawed edges inside the slot. (See Figure 3.)

Finish sand the parts of the drawer and assemble them with glue. As you clamp up the drawer, check that the assembly is square.

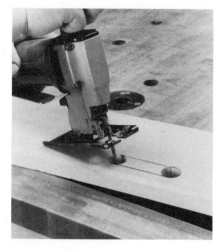

3/To cut a hand-hold in the drawer front or back, drill two 1″-diameter holes, 3″ apart, on center. Insert the blade of a saber saw in one of the holes and cut out the waste between them. If you have a Forstner bit, use it to drill the hole. Cut out the waste with a fine-tooth scrolling blade. These tools will leave a smooth surface inside the slot.

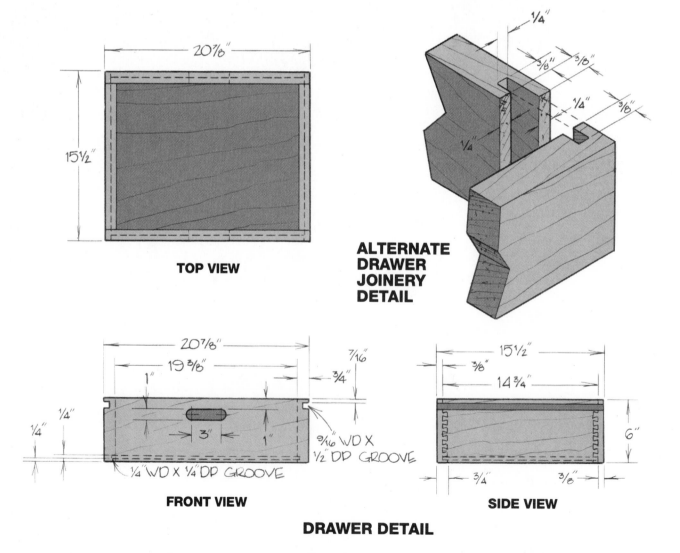

TOP VIEW

ALTERNATE DRAWER JOINERY DETAIL

FRONT VIEW

¼″ WD X ¼″ DP GROOVE

9/16″ WD X ½″ DP GROOVE

SIDE VIEW

DRAWER DETAIL

5 **Make the drawer guides.** Using a table-mounted router or a dado cutter, cut a ½"-wide, ½"-deep rabbet in one edge of each drawer guide, as shown in the *Drawer Guide Detail/End View*. Then cut a ⁹⁄₁₆"-wide, ½"-deep groove in each drawer side. The rabbeted lips of the drawer guides must fit the grooves in the drawers.

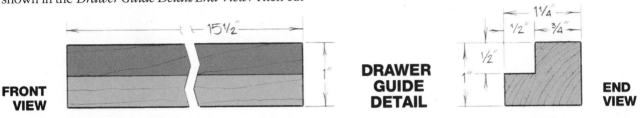

FRONT VIEW

DRAWER GUIDE DETAIL

END VIEW

6 **Cut grooves in the aprons.** Use an ordinary combination saw blade to cut ⅛"-wide, ¼"-deep grooves in the inside faces of the aprons, near the top edge. The exact distance from the top edge to the groove will depend on the tabletop clips you've purchased. These clips fit in the grooves and hold the top to the aprons, as shown in the *Top-to-Apron Joinery Detail*.

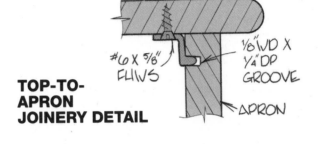

TOP-TO-APRON JOINERY DETAIL

7 **Assemble the taboret.** Finish sand all the parts of the taboret, then assemble them with glue and screws. To start, attach the drawer guides to the middle stretchers, as shown in the *Middle Stretcher/Drawer Guide Detail*. Set these parts aside. Assemble the front, back, and side aprons to make the rectangular frame to support the top. Make sure the grooves you cut earlier face the inside. In addition, these grooves should all be near the top edge of the frame. Attach the legs to the frame, then attach the stretchers to the legs. Connect the shelf between the two middle stretchers. Wherever you drive a screw, counterbore and countersink the head. Cover the heads with wooden plugs and sand them flush.

Turn the taboret assembly upside down on a flat surface. Drill holes in the bottoms of the legs to hold casters, and install them. Place the tabletop upside down on the surface, and center the taboret assembly over it. Place clips in the grooves, spacing them evenly.

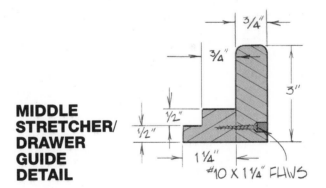

MIDDLE STRETCHER/ DRAWER GUIDE DETAIL

Attach the clips to the tabletop with screws. This arrangement will secure the top to the taboret, but still allow it to expand and contract with changes in temperature and humidity.

Finally, turn the taboret right side up and slide the drawer in place. Check that you can move the drawer in and out easily. If it binds, remove additional stock from the grooves in the sides of the drawers.

8 **Finish the taboret.** Remove the drawer and the top from the taboret, and detach the casters and the clips. Set the hardware aside. Do any necessary touch-up sanding, then apply a stain, paint, or finish to the completed project. Leave the inside of the drawer unfinished, but be sure to finish both the upper and lower surfaces of the tabletop and the shelf. This will keep these wide, flat boards from warping.

Child's Rolltop Desk

Every kid needs a desk — a special, personal work space to draw, build models, play with miniatures, or do homework. A desk also can be a place where fantasies and ambitions take shape. Having your own desk is an important step in growing up.

This child-size rolltop provides just such a place. It has a few pigeonholes and a tambour cover, with adequate work space at a comfortable height. You can easily adjust the size for different ages. There's storage for toys and other small items, and kids can roll down the top to hide the mess when parents say it's time to clean up the room. In addition, the construction is strong enough to withstand the abuse that children can dish out.

The desk is simple to build. Even with the fancy tambour top, you don't need any special skills or tools. The pigeonholes are joined with ordinary dadoes and rabbets; the larger parts are just doweled together. You can build it in a weekend!

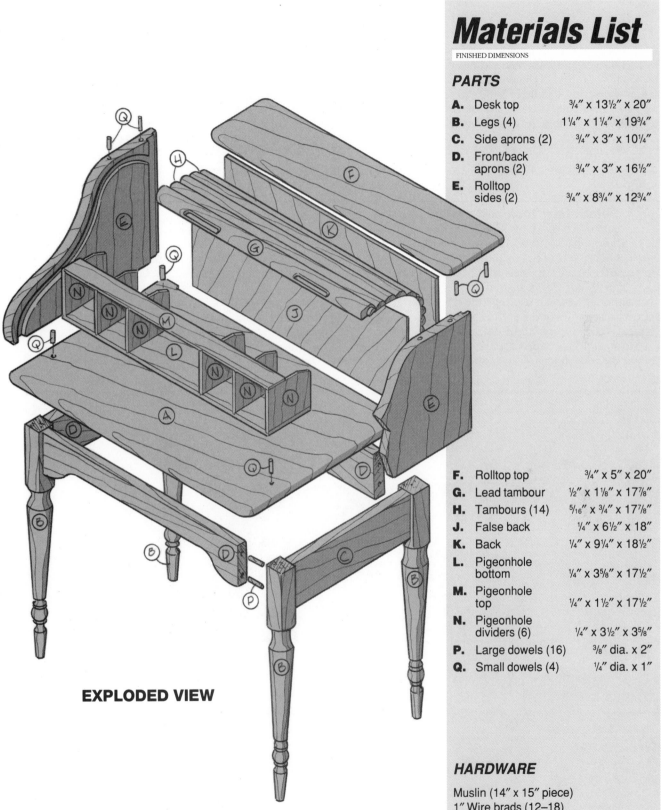

EXPLODED VIEW

Materials List

FINISHED DIMENSIONS

PARTS

A.	Desk top	¾″ x 13½″ x 20″
B.	Legs (4)	1¼″ x 1¼″ x 19¾″
C.	Side aprons (2)	¾″ x 3″ x 10¼″
D.	Front/back aprons (2)	¾″ x 3″ x 16½″
E.	Rolltop sides (2)	¾″ x 8¾″ x 12¾″

F.	Rolltop top	¾″ x 5″ x 20″
G.	Lead tambour	½″ x 1⅛″ x 17⅞″
H.	Tambours (14)	⁵⁄₁₆″ x ¾″ x 17⅞″
J.	False back	¼″ x 6½″ x 18″
K.	Back	¼″ x 9¼″ x 18½″
L.	Pigeonhole bottom	¼″ x 3⅝″ x 17½″
M.	Pigeonhole top	¼″ x 1½″ x 17½″
N.	Pigeonhole dividers (6)	¼″ x 3½″ x 3⅝″
P.	Large dowels (16)	⅜″ dia. x 2″
Q.	Small dowels (4)	¼″ dia. x 1″

HARDWARE

Muslin (14″ x 15″ piece)
1″ Wire brads (12–18)
#10 x 1¼″ Flathead wood screws (8)

1 **Adjust the dimensions to fit your child.**
As designed, this desk will accommodate children from 3 – 7 years old. If you build this project for an older child, adjust the dimensions to make the desk larger. Generally, a child of 3 – 7 should have a desk 18″–20″ tall; a child of 7 – 11 a desk 22″–24″ tall; and a child 11 – 13 a desk 26″–28″ tall. After 13, most children are comfortable with adult-size furniture. If in doubt, make the desk large — he'll grow into it.

Depending on the age and interests of the child, you may want to add more pigeonholes or a drawer under the desk top. For instructions on how to install a drawer under a work surface, see the *Contemporary/Classic Writing Table*.

2 **Select the stock and cut the parts to size.** To build this project as shown, you'll need approximately 10 board feet of 4/4 (four-quarters) stock, 3 board feet of 8/4 (eight-quarters) stock, and one quarter of a 4′ x 8′ sheet of ¼″ cabinet-grade plywood. Rolltop desks are traditionally made from oak, although you can use any kind of cabinet-grade hardwood or softwood.

Plane 8 board feet of the 4/4 stock to ¾″ thick for the sides, tambours, aprons, desk top, and rolltop top. Resaw the remaining 4/4 stock on a band saw and plane it to ¼″ thick to make the pigeonhole parts. (Use the ¼″ plywood for the false back and back.) Plane the 8/4 stock to 1¼″ thick to make the legs. When you have planed the wood to the proper thicknesses, glue up wide stock to make the desk top. Cut all parts to the sizes in the Materials List, except the legs and tambours. Cut the legs 1″–2″ long, so you have some extra stock with which to mount them on the lathe. Don't cut the tambours at all — just cut a board for tambour stock. Make this board at least 5″ wide and 17⅞″ long.

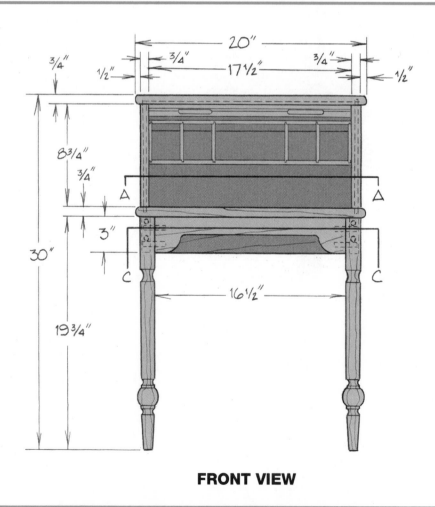

FRONT VIEW

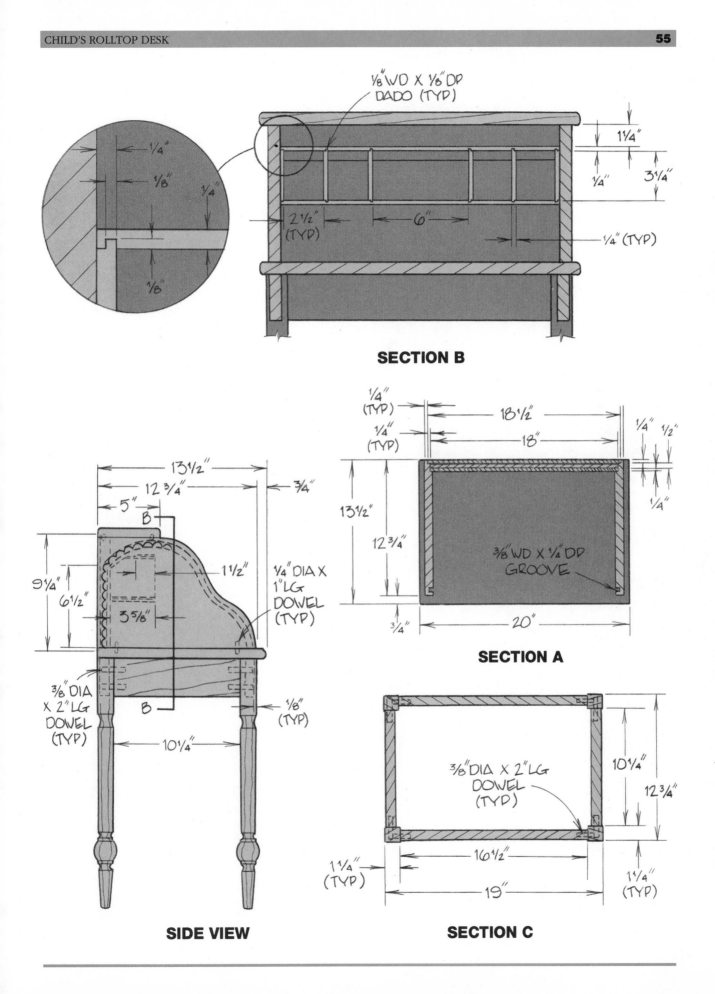

⅛" WD X ⅛" DP
DADO (TYP)

1¼"

¼"

3¼"

2½"
(TYP)

6"

¼" (TYP)

SECTION B

¼"
(TYP)

18½"

¼" ½"

¼"
(TYP)

18"

¼"

13½"

12¾"

⅜" WD X ¼" DP
GROOVE

¾"

20"

SECTION A

13½"

12¾"

5"

B

¾"

1½"

¼" DIA X
1" LG
DOWEL
(TYP)

9¼"

6½"

3⅝"

⅜" DIA
X 2" LG
DOWEL
(TYP)

B

⅛"
(TYP)

10¼"

SIDE VIEW

⅜" DIA X 2" LG
DOWEL
(TYP)

10¼"

12¾"

1¼"
(TYP)

16½"

19"

1¼"
(TYP)

SECTION C

3 Make screw pockets in the aprons.

The desk top is secured to the aprons with flathead screws. These screws rest in screw pockets, as shown in the *Side/Top/Apron Joinery Detail*. Drill these screw pockets in the inside surfaces of the aprons. (See Figures 1 and 2.) Make one pocket in each of the side aprons, centered between the ends. Make three pockets each in the front and back aprons, spaced evenly along the length.

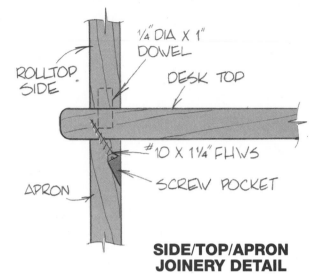

SIDE/TOP/APRON JOINERY DETAIL

1/Drill screw pockets in two steps. First, make the pocket itself: Mount a ⅝″-diameter bit in the drill press. Tilt the table to 15° and position a fence about ⅜″ away from the point where the drill touches the table. Set an apron on edge against the fence and drill a pocket in the **inside face.** Stop the drill about ½″ above the table, before breaking through the edge of the board.

2/Replace the ⅝″-diameter drill bit with a ³⁄₁₆″-diameter bit. Drill a pilot hole for the screw through the center of the pocket. The bit should exit the apron near the middle of the top edge.

4 Join the aprons and legs with dowels.

Drill ⅜″-diameter, 1″-deep holes for dowels in the ends of the aprons. Using a doweling jig to guide the drill, make two holes in each end of each board as shown in the *Front View* and *Side View*. Use dowel centers to locate matching holes in the legs, then drill them. (See Figure 3.) Dry assemble the legs and the aprons to check the fit of the dowel joints.

3/After drilling dowel holes in the aprons, mark the legs to make matching holes. To locate these holes, insert dowel centers in the ends of each apron. Measure and mark the location of the aprons on the legs. Then press the end of each apron against its leg. The dowel centers will leave small indentations, showing where to drill the holes.

5 **Turn the shapes of the legs.** Turn each of the legs on a lathe, trying to match them as closely as possible. If you have a lathe duplicating jig, use it to make the legs. If not, follow the procedure in Step-by-Step: Making Duplicate Turnings. Finish sand the legs on the lathe, and cut them to length.

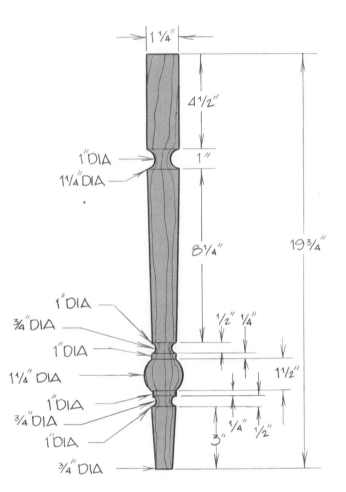

LEG LAYOUT

6 **Cut the rabbets and dadoes.** Several of the parts in the rolltop section must be dadoed or rabbeted. Make these joints with either a dado cutter or a router and straight bit. Here's a list of the joints and parts to cut:
- ¼″-wide, ½″-deep rabbet in the back edge of each side
- ¾″-wide, ¼″-deep rabbet in the back edge of each side. This will create two stepped rabbets, as shown in the *Back Corner Detail*. (See Figure 4.)
- ¼″-wide, ¼″-deep double-blind rabbet in the back edge of both the desk top and the rolltop top, as shown in the *Desktop Layout* and *Rolltop Top Layout*

4/Cut two rabbets in the back edge of each side to form steps, as shown. This first step (closest to the front edge) will hold the false back, and the second will hold the back.

- ½″-wide, ¼″-deep double-blind rabbets in the bottom edge of the lead tambour, as shown in the *Lead Tambour Layout and Profile*. These rabbets will serve as handholds. (See Figure 5.)
- ¼″-wide, ¼″-deep rabbet in each end of the lead tambour, as shown in the *Lead Tambour Layout*
- ⅛″-wide, ⅛″-deep rabbets in the top and bottom edges of the pigeonhole dividers, as shown in the *Pigeonhole Divider Layout*
- ⅛″-wide, ⅛″-deep dadoes in the pigeonhole top and bottom, as shown in *Section B*

5/Rout two blind rabbets, each 3″ long, in the bottom edge of the lead tambour. These will serve as handholds to open the tambour top.

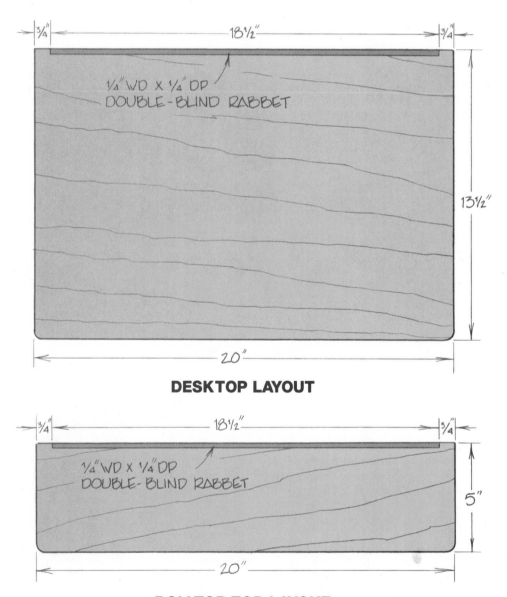

DESKTOP LAYOUT

ROLLTOP TOP LAYOUT
BOTTOM LAYOUT

7 Cut the shapes of the aprons and sides.

Enlarge the *Front Apron Pattern* and *Side Pattern*. Trace the patterns on the stock, then cut out the shapes with a band saw or saber saw. Also, cut the upper back corners of the pigeonhole dividers, as shown in the *Pigeonhole Divider Layout*. Sand the sawed edges smooth.

TRY THIS! Before you cut the sides, fasten the pieces together face-to-face with 2 or 3 wire brads. Cut and sand them both while they're fastened together. Do the same for the pigeonhole dividers — stack the six pieces and make them all at once. This saves time and ensures that all parts are precisely the same.

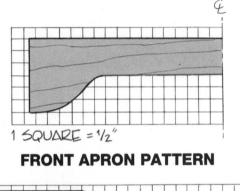

1 SQUARE = ½″

FRONT APRON PATTERN

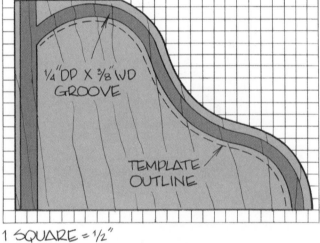

¼″ DP X ⅜″ WD GROOVE

TEMPLATE OUTLINE

1 SQUARE = ½″

SIDE PATTERN

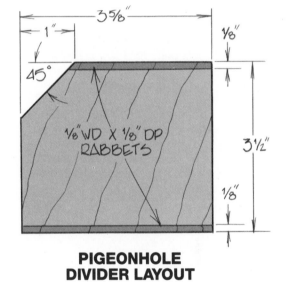

3⅝″

1″

45°

⅛″ WD X ⅛″ DP RABBETS

⅛″

3½″

⅛″

PIGEONHOLE DIVIDER LAYOUT

8 Cut the tambour grooves in the sides.

To rout the grooves for the tambours, first make a template to guide the router. Do this by tracing the enlarged *Side Pattern* onto a scrap of ½″ plywood or particleboard, including the dotted outline of the template. (The template outline is inset precisely ⅛″ from the inside edge of the groove you want to make.) Cut the template shape with a band saw or saber saw, and sand the sawed edges.

Attach the template to the inside surface of one side with wire brads. Mount a ⅜″ straight bit in the router and a ⅝″ guide bushing to the router base. (See Figure 6.) Adjust the depth of cut so the router will cut a groove ¼″ deep in the side.

6/To rout the tambour grooves, use a template to guide the router. Attach a guide bushing to the base of the router to rub against the template, then mount a straight bit so it protrudes through the bushing.

Place the router on the template and rout the groove, keeping the guide bushing pressed against the template edge. (See Figure 7.) When you've finished, remove the template from the rolltop side, turn it over, and attach it to the other side. Rout the second groove, following the same procedure.

7/Attach the template to the stock, then rout the groove, keeping the guide bushing pressed against the edge of the template.

9

Round the edges of the desk top, rolltop top, and false back. Remove the guide bushing from the router base, and mount a ⅜″ quarter-round bit in the router. This bit must have a *pilot bearing*. Round over the front and side edges of both the desk top and the rolltop top. Sand the edges to remove any machine marks.

Also, round the upper edge of the false back with a file, as shown in the *Back Corner Detail*. It should match the curve of the groove where the false back joins the side.

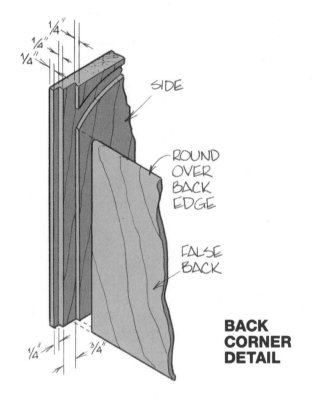

BACK CORNER DETAIL

10

Make the tambour lid. While the quarter-round bit is still mounted in the router, round the top edge of the lead tambour, as shown in the *Lead Tambour Profile*. Using a file, carefully round the rabbeted ends as shown in the profile.

Make the tambours from the 5″ x 17⅞″ tambour board you cut earlier. Using the quarter-round bit again, rout the tambour profile in one edge of the board, then rip the rounded edge free. (See Figure 8.) Repeat until you've made 14 straight, clear tambours, with no warps or twists.

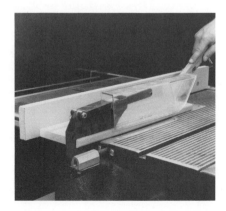

8/Round over the edge of the tambour stock, then rip a tambour free from the board. Don't try to shape the tambours after you've ripped them; the narrow stock may come apart in your hands.

Finish sand all the tambours. Cut a scrap of ¾″ plywood, 13″ x 18″ — slightly larger than the finished tambour lid will be. Lay the plywood on your workbench and cover it with waxed paper, then put a strip of muslin, 15″ wide and 14″ long, over the waxed paper. Bend the muslin around the edges of the plywood and hold it in place with thumbtacks. Cover the muslin with glue. Place the lead tambour to the extreme right or left of the muslin, then arrange the other tambours next to it edge to edge. Lay strips of scrap wood on top of the tambours, perpendicular to them, and use them to clamp the tambours to the muslin. (See Figure 9.)

When the glue dries, remove the clamps and the tacks, then peel the tambour lid off the waxed paper. With a sharp knife or a razor blade, carefully trim the excess

9/Glue the tambour to the muslin backing. Use a scrap of plywood and strips of hardwood to hold the tambours in place until the glue dries.

muslin from the top and bottom of the completed lid. Flex the lid several times to loosen the glue-stiffened muslin.

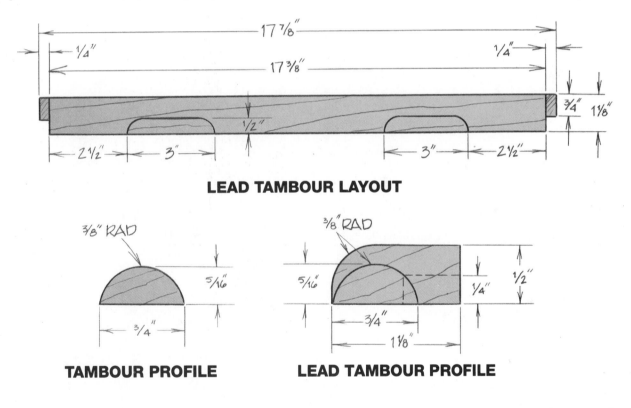

LEAD TAMBOUR LAYOUT

TAMBOUR PROFILE

LEAD TAMBOUR PROFILE

---

11 **Assemble the desk.** Finish sand the parts of the desk that you haven't sanded already. Attach the legs to the aprons with glue and dowels, making sure that the assembly is square as you clamp it together. When the glue dries, put the desk top upside down on a flat surface. Place the leg/apron assembly on the desk top. Align the back apron flush with the back edge of the desk top, and center the side aprons side-to-side. Screw the aprons to the top, driving the screws through the screw pockets. Set the assembly right side up.

Glue together the pigeonhole top, bottom, and dividers. Once again, make sure this assembly is square. When these parts have dried, glue together the pigeonholes, sides, and false back. Drill ¼″-diameter, ½″-deep holes in the top edge of the sides. Put dowel centers in these holes, and put the rolltop top in place. Press the top down so the centers leave indentations, then drill matching holes in the underside of the rolltop. Attach the top to the assembly with glue and dowels.

Attach the rolltop to the desk top in the same manner. Drill ⅜"-diameter, 1"-deep holes in the sides, put dowel centers in the holes, and press the rolltop down on the desk top. Drill matching holes where the centers leave indentations, and join the parts with glue and dowels.

Slide the tambour lid into the tambour grooves from the back. Test the rolltop action, sliding the lid up and down. If it binds, you may have to remove a little stock from the edges of the tambours. When you're satisfied that the lid slides smoothly, remove the lid from the grooves.

12 **Finish the rolltop desk.** Do any necessary touch-up sanding, then apply an equal amount of finish to all wooden surfaces, top and bottom, inside and outside — this will keep the desk from warping. However, be careful not to let the finish build up in the tambour grooves or between the tambours

themselves. The lid may stick. When the finish dries, rub it smooth and buff it with wax.

Put the tambour lid in the grooves, then attach the back to the desk with wire brads. *Don't* glue the back in place. You — or your children — may want to remove it someday to repair the lid.

Step-by-Step: *Making Duplicate Turnings*

When you make a desk, a table, or a chair with turned legs, you must *duplicate* the turnings. All four legs must match. This sounds difficult to do on a simple lathe. Without an expensive lathe duplicator,

you can't simply measure and cut one spindle after another the way you replicate wooden parts on other tools. There is, however, a simple method for making duplicate turnings without additional equipment.

1

The first step in making duplicate turnings is to trace your pattern on a **storystick** — a thin scrap of plywood or hardboard that you can hold up to the spindles as you turn them. Put all the **major** and **minor diameters** of the turning on the stick. Major diameters are the crests of the beads, diameters of flats, and the largest parts of tapers. Minor diameters are the bottoms of coves and the narrowest parts of tapers.

2

Mount the stock on the lathe and, before you begin turning, use the storystick to mark any **shoulders** — those areas where the spindle breaks from a square shape to round.

Note: Each time you use the storystick, line it up with the same reference point on the spindle. You may want to make a special reference mark.

Step-by-Step: Making Duplicate Turnings — Continued

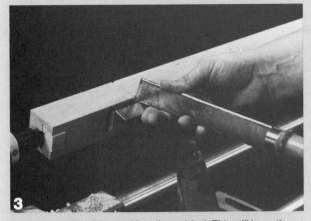

3 **Score the shoulders** with a skew chisel. This will keep the corners of the square sections from chipping or splintering as you work. Make sure your chisel is razor sharp!

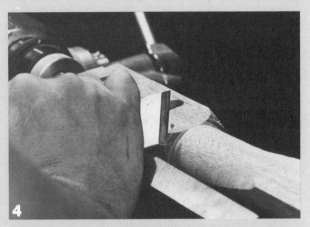

4 **Turn the stock below the shoulders,** creating a cylinder. Do **not** turn any beads, coves, or other shapes at this time. However, if the spindle design calls for it, you may round over the corners of the shoulders.

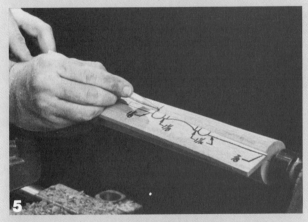

5 **After you've turned the cylinder,** use the storystick to mark the position of the beads, coves, flats, and tapers.

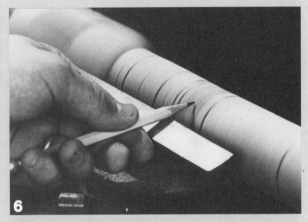

6 **Turn on the lathe at a slow speed** and darken the pencil marks so they stand out. Mark the transition between beads, coves, and other shapes with a thick line. Mark the crests of beads and other major diameters with a thin line. Lightly shade between two lines to indicate a cove. These marks will help you identify the contours as you work, without constantly referring to your plans.

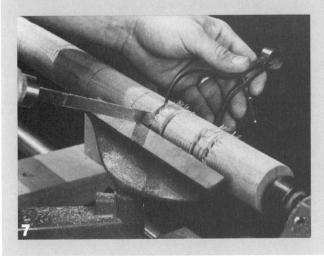

7 **When you have marked the spindle,** turn the major and minor diameters with a parting tool. Set a calipers to a particular diameter. Hold the calipers in one hand, and the parting tool in the other. Turn on the lathe and feed the parting tool into the wood. Let the calipers rest in the groove created by the parting tool. As soon as the calipers slip over the spindle, stop cutting. Repeat this procedure for each major and minor diameter, all along the length of the spindle.

(Continued)

Step-by-Step: Making Duplicate Turnings — Continued

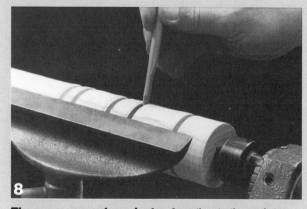

The grooves you have just cut are the starting and stopping points for each shape. As long as you turn to the bottoms of these grooves — and no further — the diameters of the various parts of the spindle will match from turning to turning. To make sure you don't cut too deeply, shade the bottom of the grooves with a pencil. As you make the beads, coves, and other shapes, be careful not to remove these pencil marks.

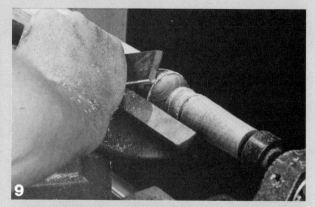

Always work from a major to a minor diameter when creating a shape. Most woodworkers prefer to make the beads and flats first: Using a skew chisel, cut from the crest of each bead down to the low spots. Cut the flats with a parting tool or straight chisel.

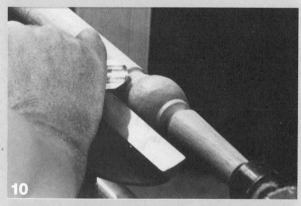

Turn the coves with a gouge, working from the high spots (the edges of the cove) to the low spots — major to minor diameter. If the spindle begins to whip or bow as you remove more and more stock, support the middle of the turning with a steadyrest.

Finally, turn the tapers. If the spindle design includes a long taper, you may want to turn the diameter of this taper every 3"–4" with a parting tool. Rough out the taper with a gouge, then finish it with a skew chisel.

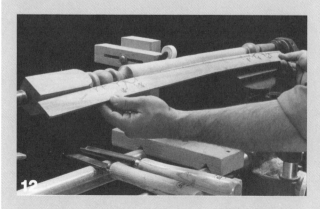

As the spindle takes shape on the lathe, check your progress now and then with the storystick. Compare the shapes on the pattern with those in the wood, and try to match them as closely as possible. By using the storystick as a guide and following this procedure for each spindle, you can match the spindles almost exactly.

Barrister's Bookcase

The stacking barrister's bookcase was one of the earliest examples of knock-down, modular furniture — and one of the most popular. It was originally designed for lawyers, to accommodate their ever-expanding legal libraries. But the public soon found that this modular design would hold almost every home and business library, no matter what sorts of books it contained.

The cases consisted of sections which stacked on top of each other. The sections came in several sizes which could be stacked in endless combinations.

"These bookcases are so constructed," read an advertisement in the 1902 Sears & Roebuck catalog introducing the concept to the general public, "that they can be adapted to the smallest library, or increased to accommodate any number of books...." Sears sold each section for under three dollars. Today, a four- or five-section antique barrister's bookcase may sell for well over a thousand dollars!

You can build this reproduction for a good deal less. It's surprisingly simple to build, too. There are only a dozen parts in each bookcase section, and most of these are assembled with ordinary rabbets, grooves, and dadoes.

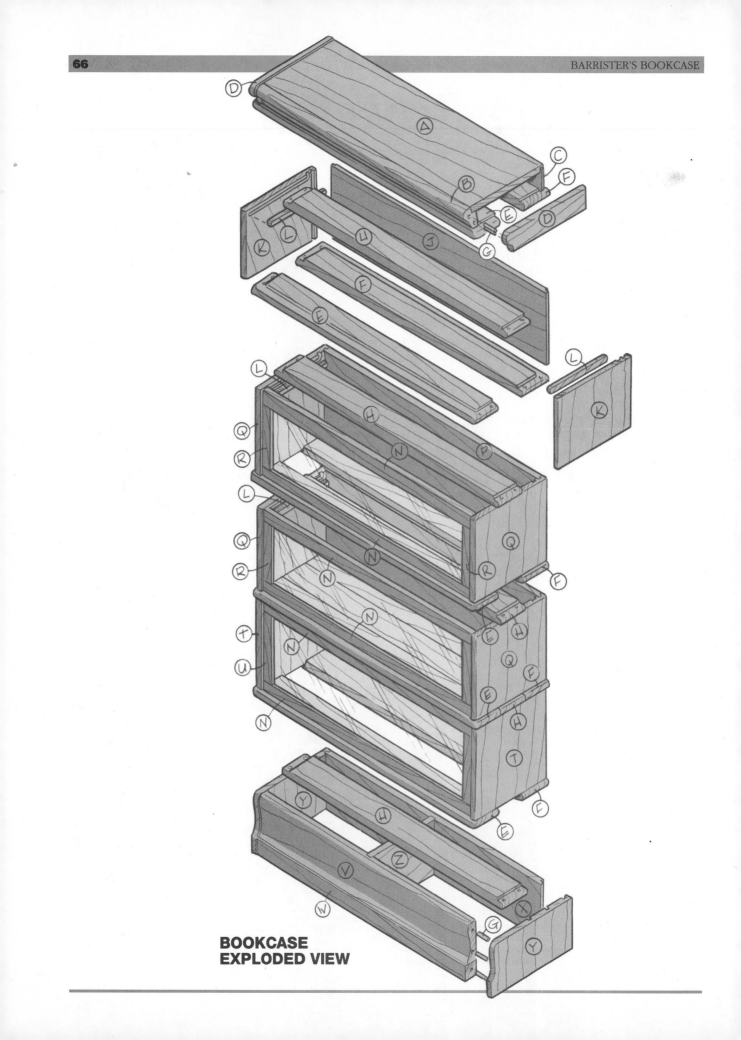

**BOOKCASE
EXPLODED VIEW**

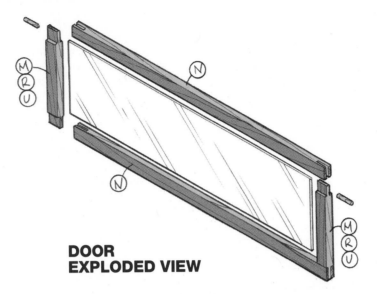

**DOOR
EXPLODED VIEW**

Materials List

FINISHED DIMENSIONS

PARTS

Top Section

A.	Top	¾" x 11" x 32½"
B.	Top molding	1½" x 2½" x 32½"
C.	Top back	¾" x 2⅛" x 33¼"
D.	Top sides (2)	¾" x 2¾" x 12¼"
E.	Front divider	1" x 4" x 34½"
F.	Back divider	1" x 3¾" x 34½"
G.	Dowels (4)	⅜" dia. x 1"

Short Bookcase Section

H.	Middle divider	1" x 3¾" x 34½"
J.	Short back	¾" x 9½" x 33¼"
K.	Short sides (2)	¾" x 9½" x 11¼"
E.	Front divider	1" x 4" x 34½"
F.	Back divider	1" x 3¾" x 34½"
L.	Cleats (2)	¾" x ¾" x 9"
M.	Short door stiles	¾" x 1½" x 9⅜"
N.	Door rails	¾" x 1½" x 32⅜"

Medium Bookcase Section

H.	Middle divider	1" x 3¾" x 34½"
P.	Medium back	¾" x 11½" x 33¼"
Q.	Medium sides (2)	¾" x 11½" x 11¼"

E.	Front divider	1" x 4" x 34½"
F.	Back divider	1" x 3¾" x 34½"
L.	Cleats (2)	¾" x ¾" x 9"
R.	Medium door stiles	¾" x 1½" x 11⅜"
N.	Door rails	¾" x 1½" x 32⅜"

Tall Bookcase Section

H.	Middle divider	1" x 3¾" x 34½"
S.	Tall back	¾" x 13½" x 33¼"
T.	Tall sides (2)	¾" x 13½" x 11¼"
E.	Front divider	1" x 4" x 34½"
F.	Back divider	1" x 3¾" x 34½"
L.	Cleats (2)	¾" x ¾" x 9"
U.	Tall door stiles	¾" x 1½" x 13⅜"
N.	Door rails	¾" x 1½" x 32⅜"

Base Section

H.	Middle divider	1" x 3¾" x 34½"
V.	Base molding	1¾" x 5" x 32½"
W.	Baseboard	2" x 2" x 32½"
X.	Base back	¾" x 7" x 33¼"
Y.	Base sides (2)	¾" x 7" x 12¼"
Z.	Brace	¾" x 7" x 10"
G.	Dowels (6)	⅜" dia. x 1"

HARDWARE

Top Section

#10 x 1¼" Flathead wood screws (12–16)

Short Bookcase Section

#10 x 1¼" Flathead wood screws (20–24)
⁵⁄₁₆" dia. x 1¼" Steel pins (2)
⅛" Glass pane (7" x 30")
Glazing points (6–8)
Door pull

Medium Bookcase Section

#10 x 1¼" Flathead wood screws (20–24)
⁵⁄₁₆" dia. x 1¼" Steel pins (2)
⅛" Glass pane (9" x 30")
Glazing points (6–8)
Door pull

Tall Bookcase Section

#10 x 1¼" Flathead wood screws (20–24)
⁵⁄₁₆" dia. x 1¼" Steel pins (2)
⅛" Glass pane (11" x 30")
Glazing points (6–8)
Door pull

Base Section

#10 x 1¼" Flathead wood screws (6–10)

1

Select the stock and cut the parts. The amount of wood needed to build this bookcase will depend on the size and number of sections you make. Each section has its own requirements:

Top section
- 4 board feet of 4/4 stock
- 3 board feet of 5/4 stock
- 2 board feet of 8/4 stock

Short bookcase section
- 6 board feet of 4/4 stock
- 4 board feet of 5/4 stock

Medium bookcase section
- 7 board feet of 4/4 stock
- 4 board feet of 5/4 stock

Tall bookcase section
- 8 board feet of 4/4 stock
- 4 board feet of 5/4 stock

Base section
- 4 board feet of 4/4 stock
- 2 board feet of 5/4 stock
- 4 board feet of 8/4 stock

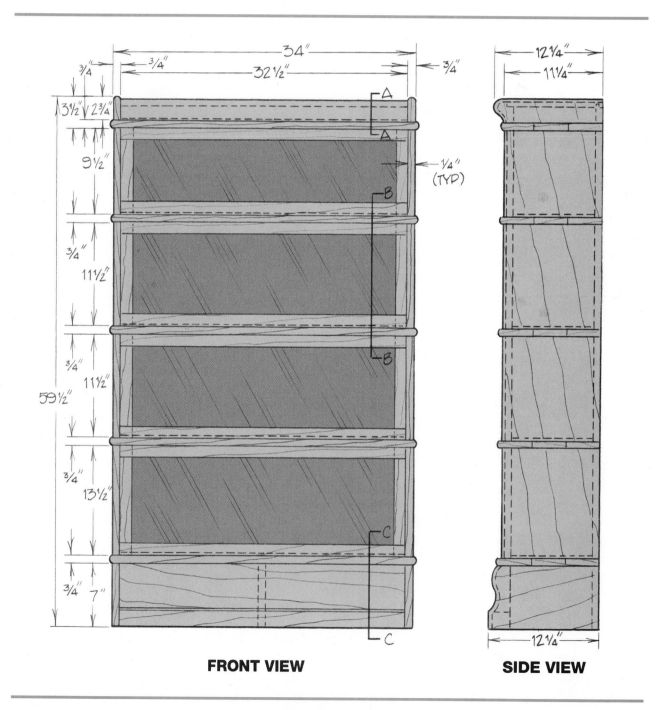

FRONT VIEW

SIDE VIEW

Make the moldings and baseboards from 8/4 (eight-quarters) stock, the dividers from 5/4 (five-quarters) stock, and all the remaining parts from 4/4 (four-quarters) stock. Traditionally, barrister's bookcases were made from oak, but you can use almost any cabinet-grade hardwood.

Plane the 4/4 stock to ¾" thick, the 5/4 stock to 1" thick, and the 8/4 stock to the various thicknesses needed to make the moldings and baseboard. Cut all parts to the sizes shown in the Materials List. Carefully label each part so you know which section it belongs to, and keep the parts in separate stacks as you work.

TRY THIS! For a striking reproduction bookcase, purchase *quarter-sawn* white oak to make the sides, stiles, rails, and moldings. Turn-of-the-century furniture manufacturers used quarter-sawn oak for its distinctive grain pattern — and because it was extremely stable. Owing to the manner in which it's cut from the log, quarter-sawn lumber is less likely to warp or cup than ordinary lumber.

2 Cut the rabbets, dadoes, and grooves needed.

Most of the parts are assembled with simple rabbets and dadoes. Make these joints with a dado cutter or a router and a straight bit. Here's a list of the joints and the parts to cut:

- 1¼"-wide, ¼"-deep rabbet in the front edge of each front divider, as shown in the *Divider Layout*
- 1"-wide, ¼"-deep rabbets in the ends of the front, middle, and back dividers
- ¾"-wide, ¼"-deep rabbet in the back edge of each back divider
- ¾"-wide, ⅜"-deep dadoes centered in the base molding, baseboard, and base back, as shown in the *Base/Top View*
- ½"-wide, ¾"-deep rabbet in the back edge of the top molding as shown in the *Top Side Pattern and Top Molding Profile*
- ⅜"-wide, ⅜"-deep groove, ⅜" from the back edge of each bookcase and base side, as shown in the *Side Layout* and *Base Side Pattern and Base Molding Profile*
- ⅜"-wide, ⅜"-deep rabbet in each end of each back
- ⅜"-wide, ⅜"-deep rabbet in the top edge of the top back, as shown in *Section A*
- ⅜"-wide, ⅜"-deep groove, ⅜" from the back edge of the top, as shown in *Section A*
- ⅜"-wide, ⅜"-deep blind groove, ⅜" from the back edge of each top side, as shown in the *Top Side Pattern and Top Molding Profile*

Note: It's easier to use a router to make the blind grooves. Start cutting at the open end, then stop before you cut through to the opposite side of the board. Square the blind end with a chisel. Use a straightedge to guide the router.

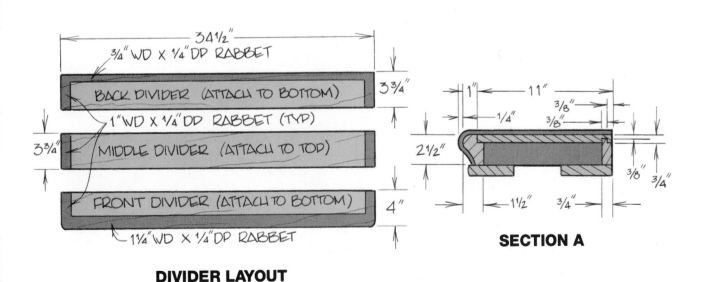

DIVIDER LAYOUT

SECTION A

3 Cut the dadoes that hold the doors.

The bookcase doors are mounted in ⅜″-wide, ¼″-deep blind dadoes in the sides. These dadoes allow you to open the doors and slide them back into the bookcase. The dadoes are open at the back end so you can remove the doors should they need repair.

Cut these dadoes with a router. Use the same method you used to make the blind grooves in the top sides, with one exception: Cut the grooves in an L-shape.

To do this, first mark the dado on each side. Clamp a straightedge to the side, parallel to the dado, to guide the router. Mount a ⅜″ straight bit in the router, and hold the tool so the bit is near the top edge of the side — right where you want to cut the open end of the dado. Make sure the bit is not touching the board, then turn on the router. Carefully advance the router, moving it *perpendicular* to the top edge. Cut through the top edge until the base touches the straightedge. Then cut *parallel* to the top edge, following the straightedge. Stop cutting when the bit is ⅝″ from the front edge of the side. (See Figures 1 through 3.)

2/Change the cutting direction when the router reaches the straightedge. Turn 90˚ and cut a blind dado 9⅞″ long. Stop cutting ⅝″ before the front edge.

1/To cut the dado that holds the door in a bookcase side, first cut the open end in the top edge. This opening should be ¾″ from the back edge, as shown on the **Side Layout**.

3/The finished dado should be L-shaped, as shown.

4 Cut the shape of the moldings.

Both the top and the base moldings are shaped by multiple cuts with a table saw and a hand plane:

Top molding — First, cut a cove in the front face of the molding stock. Place a straightedge on the table of the saw, 45˚ off parallel to the blade. Adjust it to guide the edge of the stock over the center of the blade, and clamp the straightedge to the table. Lower the saw blade until it's cutting no more than ⅛″ deep. Feed the molding stock across the saw blade, guiding it along the straightedge. Raise the blade about ⅛″, and make another pass. Continue until the cove is ½″ deep. (See Figure 4.)

Once you've cut the cove, clamp the stock to your workbench, front face up. Using a bench plane, round the top edge of the molding, as shown in the *Top Side Pattern and Top Molding Profile*. Cut the shape evenly all along the length of the molding, using a cardboard template to check your work. (See Figures 5 and 6.)

4/To make the top molding, first make a cove cut on a table saw. Pass the wood over a 10″-diameter table saw, guiding it along a fence set at 45˚ to the blade. Make the cut in several passes, cutting just ⅛″ deeper with each pass. Feed the stock very slowly across the blade.

Base molding — Use a similar method to shape the base molding. Set up the table saw to cut a cove in the front face of the stock. Angle the straightedge at 30° to the blade and position it 1¼" back from the blade's center. Pass the wood over the blade, cutting the cove 1¼" up from the bottom edge. Round over the top edge with a bench plane, checking your work with a template. The profile of the finished molding should match the shape shown in the *Base Side Pattern and Base Molding Profile.*

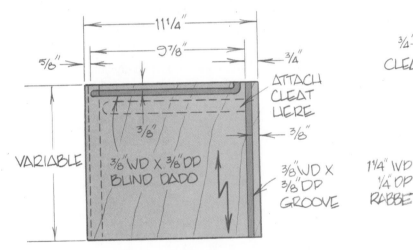

5/Round the top edge of the molding by hand, using a bench plane. Take long strokes, running the plane down the entire length of the stock with each stroke. This will help keep the shape even.

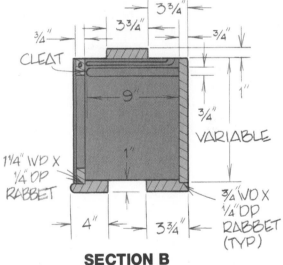

6/As the molding begins to take shape, check your work with a cardboard template. The template must fit the molding all along its length.

SIDE LAYOUT

11¼"
9⅞"
5/8"
3/4"
ATTACH CLEAT HERE
3/8"
3/8"
VARIABLE
3/8" WD X 3/8" DD BLIND DADO
3/8" WD X 3/8" DP GROOVE

SECTION B

3¾"
3¾"
3/4"
3/4"
CLEAT
9"
1"
3/4"
VARIABLE
1"
1¼" WD X ¼" DP RABBET
3/4" WD X ¼" DP RABBET (TYP)
4"
3¾"

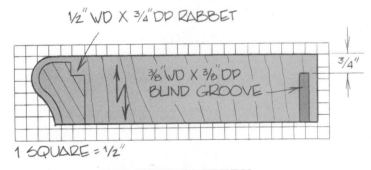

TOP SIDE PATTERN AND TOP MOLDING PROFILE

½" WD X ¾" DD RABBET
3/4"
3/8" WD X 3/8" DD BLIND GROOVE
1 SQUARE = ½"

SECTION C

3¾"
1¾"
3¾"
5"
2"
¼"
2"
2"
3/4"
12"

5 Cut the shapes of the sides and cleats.

Enlarge the *Top Side Pattern* and *Base Side Pattern* and trace them on the stock. These pieces should extend exactly ¼″ beyond the front edge of the moldings. If a molding shape has changed slightly from that shown in the drawings, adjust the shape of the side to match it. Cut the shapes with a band saw or saber saw, and sand the sawed edges.

Also, round the front ends of the cleats. With a compass or a circle template, draw a ⅜″ radius on the end of each cleat. Using a disk sander or a belt sander, sand up to the radius mark, rounding the end.

6 Round the protruding ends and edges.

The ends and edges of several parts of this bookcase protrude slightly from other parts. For example, the top sides protrude ¼″ above the top. The dividers protrude ¼″ beyond the bookcase sides. The bookcase sides protrude ¼″ in front of the doors, and so on. Using a router and a ½″ quarter-round bit, round over all the protruding ends and edges, as shown in the *Edge Detail*. Sand the rounded areas to remove the mill marks.

7 Assemble the top and base sections.

Drill ⅜″-diameter, ½″-deep dowel holes in the ends of the moldings and baseboard. Carefully mark where these parts will join the sides. Put dowel centers in the dowel holes, and press each part against the side where it will be joined. The dowel centers will leave small indentations. Drill matching dowel holes in the sides at each of these marks.

Finish sand all the parts of the top and base sections. Then assemble the parts with glue, dowels, and screws in this order:

Top section –– Glue dowels in the ends of the top molding. Assemble the top molding, sides, and back with glue. Then glue the top to the top molding and the back. Attach the front and back dividers to the assembly with screws, but don't glue them yet. It's best to wait until you've assembled the entire bookcase before gluing them permanently in place.

Base section — Glue dowels in the ends of the baseboard and base molding, then glue the baseboard and the base molding together. Before the glue dries completely, glue the base molding, baseboard, sides, and back together. Then slide the brace in position. Check that the assembly is square as you clamp it up. Let the glue dry, then screw (but *don't* glue) the middle divider to the sides and brace.

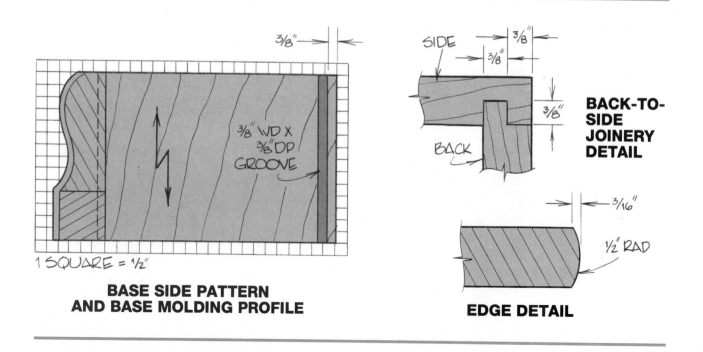

**BASE SIDE PATTERN
AND BASE MOLDING PROFILE**

BACK-TO-SIDE JOINERY DETAIL

EDGE DETAIL

8 Assemble the bookcase sections.

Finish sand all the bookcase parts. Drill ¼″-diameter pilot holes in the cleats, then attach the cleats to the inside surfaces of the sides with screws. The pilot holes are slightly oversize to allow the sides to expand and contract. *Don't* glue the cleats in place; the glue will restrict the movement of the sides.

Glue the sides to the back. Then screw (but *don't* glue) the dividers to the top and bottom edges of the sides. **Warning:** When you attach the middle divider, make sure the screws don't go through the grooves in the sides. You may want to use finishing nails instead of screws.

9 Glue the dividers in place.

Stack the sections to make sure they all fit together properly. If they don't, you may want to change the position of the dividers slightly. When all the sections fit, remove the *front* and *back* dividers from the top and bookcase sections. Glue them back to their respective sections, and replace the screws. Do the same for the middle divider on the base section.

Do not glue the middle dividers to the bookcase sections. You'll need to remove these to install the doors.

They should remain removable in case you ever need to repair the doors.

TRY THIS! To move a divider slightly one way or the other, shave a matchstick so it fits in the pilot hole. Glue the matchstick in the hole, then drill a new pilot hole beside the old one.

10 Make the doors.

Cut the bridle joints (sometimes called slot mortises or open mortises and tenons) that join the parts of the door frames. Using a tenon-cutting jig, make tenons in the ends of the door rails and mortises in the ends of the door stiles, as shown in the *Door Frame Joinery Detail*. (See Figure 6.) Assemble the door frames with glue, checking that the frames are absolutely square as you clamp them together.

7/Use a tenon-cutting jig to cut both the mortises and the tenons of the bridle joints. The jig shown can be easily made from scraps of wood. It rides along the table saw fence, guiding the wood over the blade.

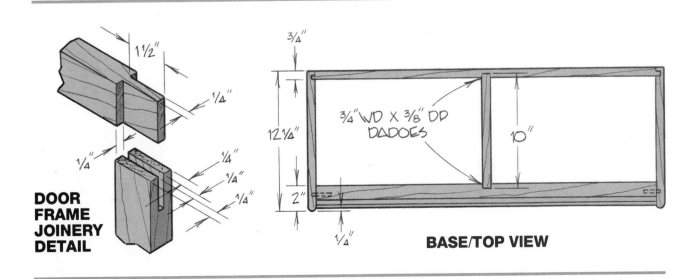

DOOR FRAME JOINERY DETAIL

1½″
¼″
¼″
¼″
¼″
¼″
¼″

¾″
12¼″
2″
¼″

¾″ WD X ⅜″ DP DADOES

10″

BASE/TOP VIEW

Note: The door frame stock must be perfectly straight, with no twists, warps, or other defects. *This is important!* If the frame parts are even slightly distorted, the doors will not fit the case correctly. Also make sure the frame is flat when you assemble it.

When the glue dries on the door frames, rout the back of the frames, making 3/8"-wide, 3/8"-deep rabbets all the way around the inside edges. Square the corners of the rabbets with a chisel. Finish sand the completed frames, making all the joints clean and flush.

11 Mount the doors.

Drill 5/16"-diameter, 7/8"-deep holes near the upper corners of the door frames, as shown in the *Door/Back View*. Insert 5/16"-diameter, 1 1/4"-long steel pins in each of these holes. Install a small door pull, centered on the bottom rail of the door frame.

Remove the middle divider from each bookcase section. Put the door in place, inserting the ends of the

metal pins in the open ends of the L-shaped dadoes. Check that the door opens, shuts, and slides easily. If it binds at any point, remove stock from the edges of the door or the inside edges of the dadoes, as necessary. You may also need to move the cleats. When you're satisfied that the door operates properly, replace the middle divider.

12 Finish the completed sections.

Remove the doors from the bookcase sections, and remove the pins and pulls from the doors. Do any necessary touch-up sanding to the wooden surfaces, then apply a finish. Be sure to apply as many coats to the inside of each section as you do to the outside.

When the finish dries, rub it out and apply a coat of wax.

Reassemble the doors and the door hardware. Place panes of glass in the rabbets in the door frames. Secure the glass with glazing points. If the glass rattles when you open or close the doors, put a tiny piece of felt between the glass and each glazing point.

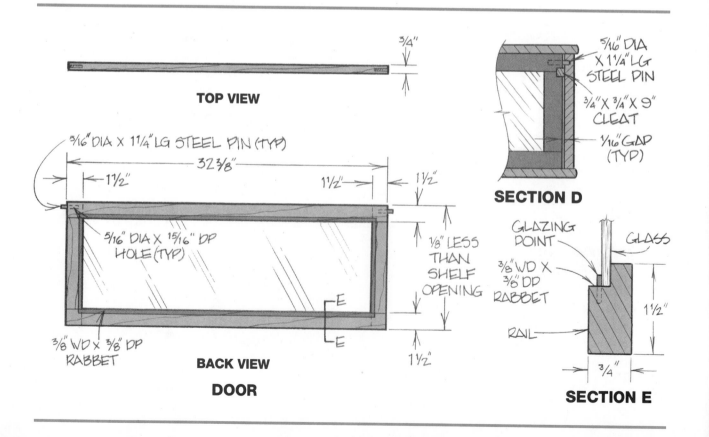

Wooden Filing Cabinet

Just as every office needs a desk, every desk needs a filing cabinet. While a desk provides a place to *do* paperwork, a filing cabinet is a place to *store* it. Even if you have just a small household, you must keep (and be able to retrieve) deeds and mortgages, cancelled checks, insurance policies, birth records, titles, warranties — enough to overload most desks. If you have more papers than you have room to put them, then it's time to build a filing cabinet.

A filing cabinet is an ordinary chest of drawers with two special features. First, the drawers are built to accommodate standard-size papers and file folders, either letter size or legal size. Second, the drawers slide all the way out of the cabinet, letting you reach the files at the back.

This filing cabinet is easier to build than most. The traditional drawer supports have been eliminated, making the case a hollow shell. A solid plywood back keeps the case square, and the drawers travel in and out on metal slides. ◉

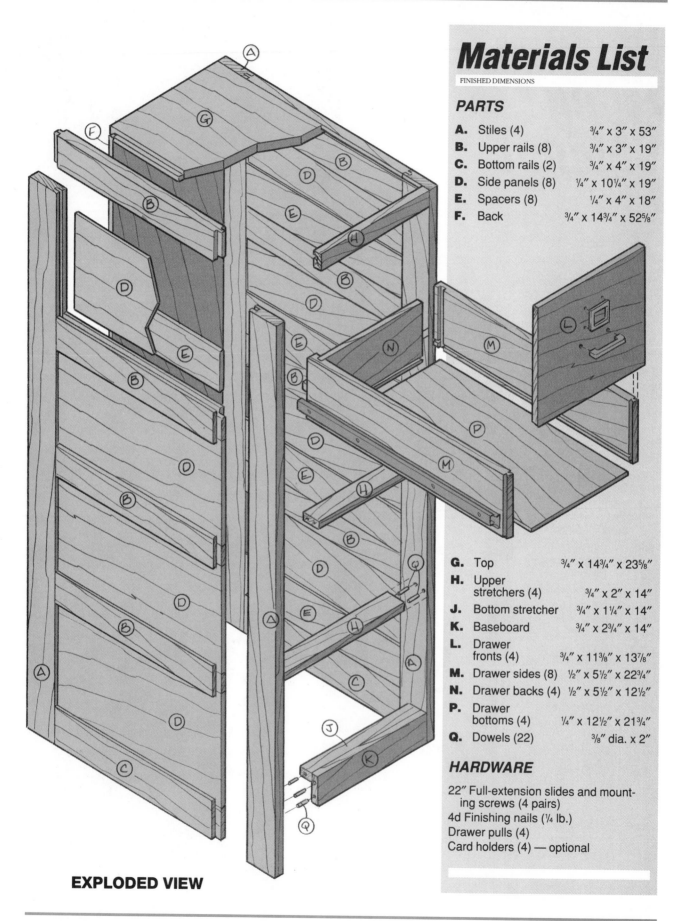

EXPLODED VIEW

Materials List
FINISHED DIMENSIONS

PARTS

A.	Stiles (4)	¾" x 3" x 53"
B.	Upper rails (8)	¾" x 3" x 19"
C.	Bottom rails (2)	¾" x 4" x 19"
D.	Side panels (8)	¼" x 10¼" x 19"
E.	Spacers (8)	¼" x 4" x 18"
F.	Back	¾" x 14¾" x 52⅝"

G.	Top	¾" x 14¾" x 23⅝"
H.	Upper stretchers (4)	¾" x 2" x 14"
J.	Bottom stretcher	¾" x 1¼" x 14"
K.	Baseboard	¾" x 2¾" x 14"
L.	Drawer fronts (4)	¾" x 11⅜" x 13⅞"
M.	Drawer sides (8)	½" x 5½" x 22¾"
N.	Drawer backs (4)	½" x 5½" x 12½"
P.	Drawer bottoms (4)	¼" x 12½" x 21¾"
Q.	Dowels (22)	⅜" dia. x 2"

HARDWARE

22" Full-extension slides and mounting screws (4 pairs)
4d Finishing nails (¼ lb.)
Drawer pulls (4)
Card holders (4) — optional

1

Adjust the size of the cabinet as needed. As shown, this filing cabinet has four drawers, all built to hold letter-size files. Depending on your needs, you may want to build a cabinet with fewer or different-size drawers. If you want to make a two-drawer filing cabinet, for example, remove two of the paneled sections from the middle of the cabinet,

keeping the bottom and top section. If you want drawers that hold legal files, widen the project 3″, making the drawers 16⅞″ wide and the cabinet 18½″ wide. To adjust the length of the drawers, change the long dimension of the top, rails, drawer sides, side panels, and drawer bottom.

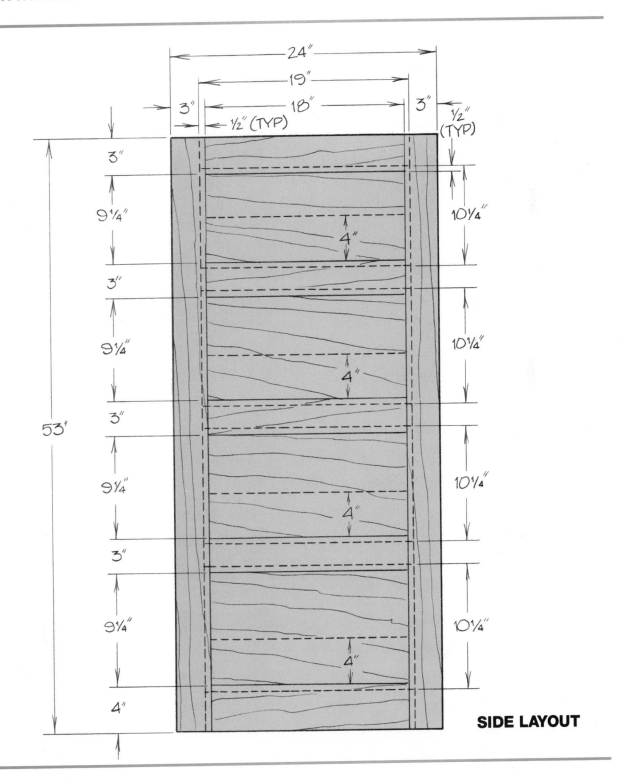

SIDE LAYOUT

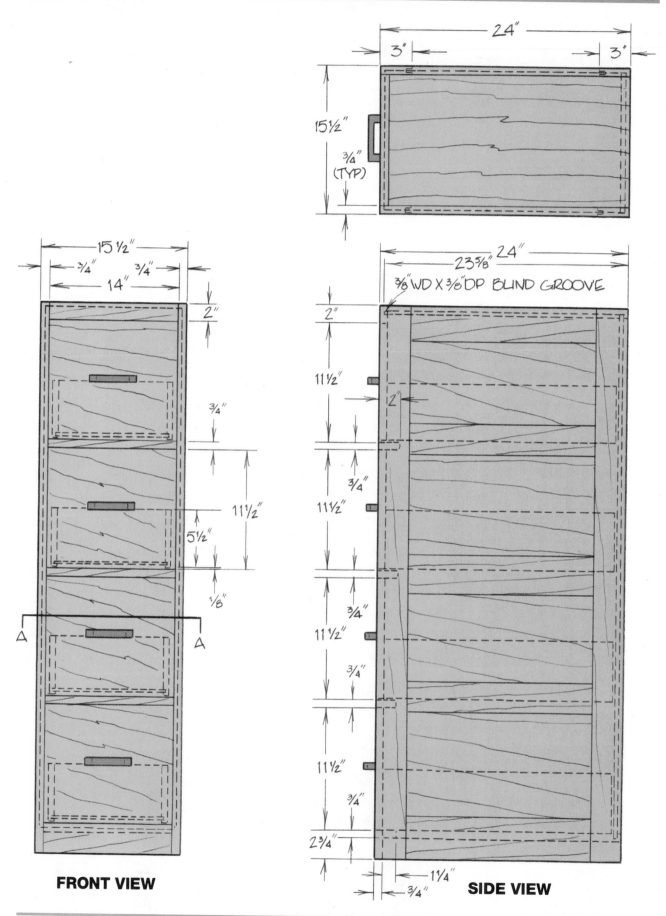

FRONT VIEW

SIDE VIEW

2 **Select the stock and cut the parts.** To make this project as shown, you need approximately 24 board feet of 4/4 (four-quarters) stock, one 4' x 8' sheet of ¾" cabinet-grade plywood, and one 4' x 8' sheet of ¼" cabinet-grade plywood. Traditionally, wooden filing cabinets are made from oak, but you can use any hardwood. Just make sure the veneer on the plywood matches the wood used for the outside parts. You can also save some money by using an inexpensive hardwood such as poplar for the drawer sides and bottoms, since these are the inside parts.

Use the ¾" plywood to make the top, back, and drawer fronts; the ¼" plywood for the side panels, spacers, and drawer bottoms; and the 4/4 solid stock for the remaining parts. Plane 12 board feet of this stock ½" thick, to make the drawer sides and backs, and the rest ¾" thick for the stiles, rails, stretchers, and baseboard. Cut the parts to the sizes given in the Materials List, except the drawer fronts. Cut these exactly ⅛" wider and longer than specified, making them 11½" by 13". *Don't* bevel the edges yet.

3 **Cut the tongues and grooves in the rails and stiles.** The sides of the filing cabinet consist of large frames and panels assembled with tongue-and-groove joinery, as shown in *Section A*. Cut these tongues and grooves in the rails and stiles, using a dado cutter or a table-mounted router and a straight bit.

First, cut ¼"-wide, ½"-deep grooves in the *inside* edges of the stiles, the *inside* edges of the top and bottom rails, and *both* edges of the middle rails. Then cut ¼"-wide, ½"-long tenons in the ends of all the rails. (See Figures 1 and 2.) The tenons must fit in the grooves with little or no slop.

1/To join the frame parts, cut grooves in the edges of the rails and stiles. If you use a dado cutter to make these grooves, guide the boards with the rip fence.

2/Cut matching tongues in the ends of the rails. Use a miter gauge to guide the stock, and clamp a stop block to the fence to set the length of the tenons.

4 **Assemble the sides.** Dry assemble the frame-and-panel sides to test the fit of the joints. When you're satisfied that the parts fit properly, finish sand the rails, stiles, and side panels. Then glue the *bottom* edge of the panels in the grooves of the rails. Before these subassemblies dry, glue the tenons of the rails into the grooves of the stiles. *Don't* glue any other part of the panels — just the bottom edge. Let the other edges float in their grooves.

As you clamp the parts together, check that the side assemblies are square. This is very important if the drawers in the filing cabinet are to work properly. Wipe away any excess glue that squeezes out of the joints with a wet rag.

When the glue dries, sand the joints clean and flush. Glue spacers to the *inside* faces of the side panels, near the bottom edges. These spacers double the thickness of the panels, providing a strong surface on which to mount the extension slides.

Note: You must use plywood for *both* the panels and the spacers, since the slides will be mounted on these parts. Plywood is stable; it doesn't expand or contract very much with changes in the weather. This, in turn, provides a stable assembly to hold the drawers.

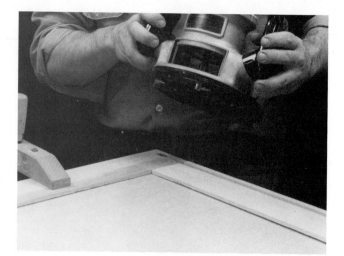

5 **Cut the dadoes and rabbets.** Most of the wooden parts are joined with ordinary rabbets, dadoes, and grooves. You can make these joints with a dado cutter or a router and a straight bit. Here's a list of the joints and parts to cut:

- ⅜″-wide, ⅜″-deep rabbets in the side edges of the back and the top, as shown in the *Back/Top/ Side Joinery Detail*
- ⅜″-wide, ⅜″-deep rabbets in the top edge of the back and front edge of the top
- ⅜″-wide, ⅜″-deep dadoes near the back edge of the top, and the top edge of the topmost upper stretcher
- ⅜″-wide, ⅜″-deep grooves near the back edges and top edges of the side assemblies

Make the grooves near the back edges of the side assemblies *blind* at the top end — stop them just before they exit the top edge. The easiest way to make a stopped groove is with a router, halting it before you cut through the edge of the wood. (See Figure 3.)

3/When routing the grooves in the larger parts and assemblies, guide the router with an edge guide acces- *sory, mounted to the base. To make the blind grooves near the back edges of the side assemblies, simply* *stop routing when you reach the other groove near the top edge. Cut through to this groove, but not any further.*

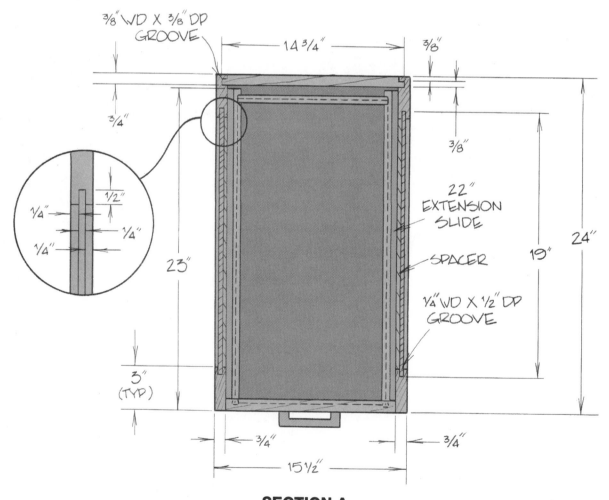

SECTION A

6

Drill the dowel joints for the stretchers and baseboard. Using a doweling jig to guide the drill, make ⅜″-diameter, 1⅜″-deep holes in the ends of the stretchers and baseboard. Drill two holes per end in the baseboard and upper stretchers; drill just one hole in each end of the bottom stretcher. Glue the bottom stretcher to the baseboard to form an L-shape, as shown in the *Stile/Stretcher Joinery Detail*. Sand the joint clean and flush.

Measure and mark the position of the stretchers on the inside surfaces of the front stiles. (Measure *twice;* you must position the stretchers precisely.) Place dowel centers in the holes in the end of a stretcher, and press the stretcher against the stile where you want to attach it. The centers will leave small indentations. Drill a ⅜″-diameter, ⅝″-deep hole at each indentation. Repeat for each stretcher and the bottom stretcher/baseboard assembly.

7

Assemble the cabinet. Finish sand the top, stretchers, and baseboard. Assemble the cabinet without glue to test the fit of the parts. When you're satisfied that they all fit properly, put them together with glue, dowels, and finishing nails. Assemble the parts in the following order.

Lay one of the side assemblies on a flat surface, so the inside faces up. Attach the baseboard and stretchers with glue and dowels. Glue the top in place, then the back. Put all the dowels in the upturned ends of the baseboard and stretchers, then glue the other side to the assembly.

Turn the assembled cabinet right side up and clamp the parts together. As you tighten the clamps, check that the cabinet is perfectly square. Let the glue dry, then

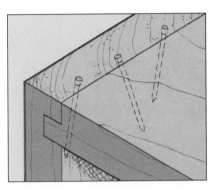

4/When using nails to reinforce the glue joints, drive the nails at slight alternating angles to help hook the parts together.

remove the clamps. Reinforce the rabbet-and-groove joints that hold the top and back to the sides with small finishing nails. (See Figure 4.) Set the heads of the nails below the surface of the wood.

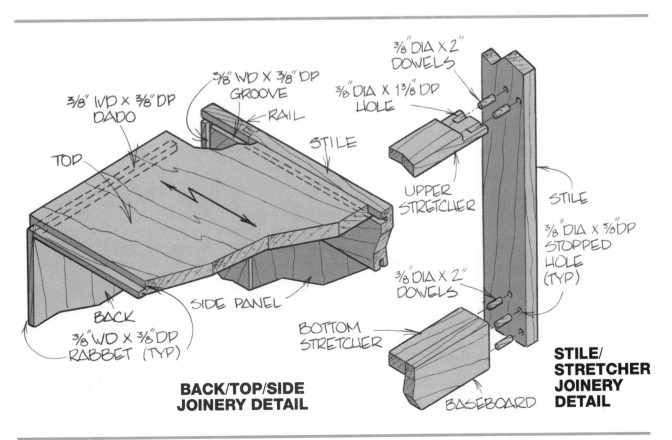

BACK/TOP/SIDE JOINERY DETAIL

STILE/ STRETCHER JOINERY DETAIL

8

Make the drawers. The drawer fronts are attached to the sides with ½″-wide, ½″-long dovetail tenons and slots. (These are sometimes called sliding dovetails or French dovetails.) The back sits in ½″-wide, ¼″-deep dadoes in the sides, and the bottom rests in ¼″-wide, ¼″-deep grooves in the front, back, and sides. Cut the dadoes and the grooves with either a dado cutter or a router and a straight bit. Make the sliding dovetails with a table-mounted router and a ½″ dovetail bit, following this procedure.

Cut the slots first, so you can fit the tenons to them. Adjust the dovetail bit ½″ above the router table, and position a fence or straightedge exactly ½″ away from the bit. Using this fence to guide the stock, cut 5¾″-long slots in the inside face of the drawer fronts. To cut matching tenons, adjust the fence and pass the drawer sides between the fence and the bit. (See Figures 5 and 6.)

5/Cut two ½″-wide, ½″-deep dovetail slots in the inside faces of each drawer front. Start at the bottom edge and rout parallel to the side edges. Clamp a stop block to the guide fence to halt the stock when each slot is 5¾″ long.

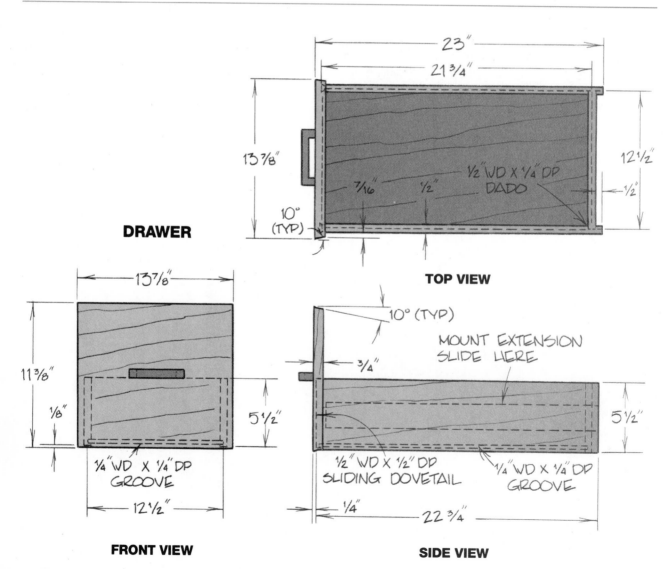

DRAWER

TOP VIEW

23″
21¾″
13⅞″
12½″
7/16″
½″
½″ WD X ¼″ DP DADO
½″
10° (TYP)

FRONT VIEW

13⅞″
11⅜″
⅛″
5½″
¼″ WD X ¼″ DP GROOVE
12½″

SIDE VIEW

10° (TYP)
MOUNT EXTENSION SLIDE HERE
¾″
5½″
½″ WD X ½″ DP SLIDING DOVETAIL
¼″ WD X ¼″ DP GROOVE
¼″
22¾″

Carefully measure the drawer openings in the front of the cabinet and cut the drawer fronts to fit. If the measurements haven't changed from what is shown in the drawings, you'll need to shave exactly ¹⁄₁₆″ off *each* edge of *each* drawer front. If they have changed (and they probably will have), remove more or less stock to compensate. Shave the drawer fronts on the table saw, beveling the edges at 10° as you do so. Drill holes in the fronts to mount the drawer pulls.

Test fit the drawer fronts in the cabinet openings — there should be a ¹⁄₁₆″ gap between the front and the cabinet, all around the perimeter of the opening. Assemble the drawers without glue to test the fit of the parts. When you're satisfied that everything fits properly, finish sand the drawer fronts. Assemble the fronts, sides, and backs with glue. Insert the drawer bottom as you assemble each drawer, but don't glue it in place. Let it float in the grooves.

Carefully check that the drawers are square as you clamp the parts together. Wipe away any excess glue that squeezes out of the joints with a wet rag.

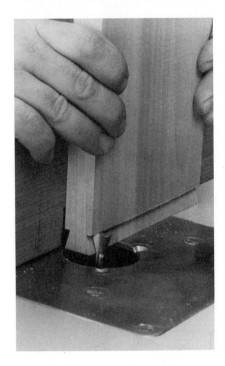

6/To cut a dovetail tenon in the end of a drawer side, adjust the guide fence so it's about ⅜″ from the bit. Pass the stock between the fence and bit, cutting one side. Turn the stock around and cut the other side. These two cuts will form a tenon. Do this in scrap stock first, fitting the tenon to the slots. If it's too loose, move the fence away from the bit. If it's too tight, move the fence closer.

9 **Hang the drawers.** Attach the drawer pulls and the optional card holders to the fronts of the drawers. Then install the extension slides in the cabinet, and attach the drawers to the slides. The exact procedure for installing a slide will vary with the manufacturer. However, it usually follows this general outline.

Most slides separate into two parts — the actual slide and a hanger or mount that mates to the slide. Attach a pair of slides to the stiles and spacers inside the cabinet, 1″–3″ above the nearest stretcher. (The exact distance will depend on the width of the slide.) Attach the mounts to the sides of a drawer, ¹⁵⁄₁₆″–2¹⁵⁄₁₆″ above the bottom edge of the drawer front.

Slip the mounts into the slides, push the drawer closed, and check the fit of the drawer front in the opening. The bottom edge should be ¹⁄₁₆″ above the stretcher. There should also be a small gap at the top and each side. If not, remove the drawer and adjust the position of the slides. The screw mounting holes in the slides are usually slotted, so you can move them up or down about ¹⁄₄″ simply by loosening the screws. Reposition the slides as necessary, and tighten the screws. Replace the drawer and check the fit again. Repeat until the drawer fits properly and slides in and out of the cabinet smoothly. Mount each drawer and adjust its position in the same manner.

10 **Finish the filing cabinet.** Remove the drawer from the completed cabinet, and detach all the hardware — slides, mounts, drawer pulls, and card holders. Do any necessary touch-up sanding, then apply a durable, easy-to-clean finish such as varnish or polyurethane. Be sure to put as many coats on the inside of the cabinet as the outside — this will keep the project from warping or distorting.

After the finish dries, polish the project with wax. Replace the hardware and the drawers.

Letter Holder

It's hard to imagine that a day's mail might fit in this antique letter holder. But in all probability, it once held a week's or even a month's worth of bills and letters.

Postmen didn't always come six days a week, particularly in hard-to-reach areas. Nor did they always stuff your mailbox with junk mail. Late into this century, some rural postmen made their rounds only once or twice a week. When they did come, they usually brought only a few letters if anything at all. It might have taken several visits to fill a small letter holder like this.

This country piece was built around the turn of this century to help organize mail for a family farm. The slots in the bracket held the letters, while the small drawer stored stamps, sealing wax (some of it remains), and perhaps a small penknife for opening envelopes. Since then, it has had several different jobs, according to the owner. The holder once served time in a country kitchen, holding recipe cards and coupons. It hung in an entrance hallway, holding telephone messages and reminders. Today, it sits in a home office, holding letters and cards from far-flung friends and family. ●

EXPLODED VIEW

Materials List

FINISHED DIMENSIONS

PARTS

A.	Back	$\frac{1}{4}$" x $4\frac{1}{2}$" x 12"
B.	Sides (2)	$\frac{1}{4}$" x 3" x $4\frac{1}{4}$"
C.	Top/bottom	$\frac{1}{4}$" x 3" x $4\frac{1}{4}$"
D.	Bracket	1" x 1" x 8"
E.	Drawer front/ back (2)	$\frac{1}{4}$" x $1\frac{15}{16}$" x $3\frac{15}{16}$"
F.	Drawer sides (2)	$\frac{1}{4}$" x $1\frac{15}{16}$" x $2\frac{7}{8}$"
G.	Drawer bottom	$\frac{1}{4}$" x $2\frac{1}{2}$" x $3\frac{7}{16}$"

HARDWARE

#8 x $\frac{3}{4}$" Flathead wood screws (2)
$\frac{3}{4}$" Wire brads (20)
Small drawer pull

1 Select the stock and cut the parts.

A small country piece like this can be made out of any domestic wood. The original craftsman probably used whatever scrap he had lying around. He built the holder shown from white pine. You might also use cherry, walnut, maple, or birch.

After choosing the stock, plane it to the thicknesses needed — $\frac{3}{4}$" for the bracket, and $\frac{1}{4}$" for the rest of the parts. Cut the parts to size, except for the drawer parts. Make these slightly longer and wider than shown in the Materials List, so the assembled drawer will be slightly oversize. Later, you can sand it to fit.

2

Cut the joinery. Using a dado cutter or a table-mounted router, cut ¼″-wide, ⅛″-deep dadoes and rabbets in the sides, as shown in the *Front View* and *Side View*. Adjust the depth of cut, then make ¼″-wide, 3/16″-deep rabbets in the ends of the drawer front and back, as shown in the *Drawer/Top View*.

3

Cut the slots in the bracket. Adjust the blade tilt of your table saw or radial arm saw to 45°. If you're using a table saw, adjust the depth of cut so the blade protrudes ½″ above the table surface. If you're working with a radial arm saw, adjust it so the blade is suspended ¼″ above the surface.

Using an ordinary combination blade, cut four angled slots in the bracket. (See Figure 1.) Space the slots evenly along the length of the bracket, as shown in the *Side View*.

*1/Cut 45° slots in the bracket, evenly spaced along the length, to hold the mail. To handle this small part on a table saw safely, attach a miter gauge extension to the miter gauge. Saw through the bracket **and** the extension.*

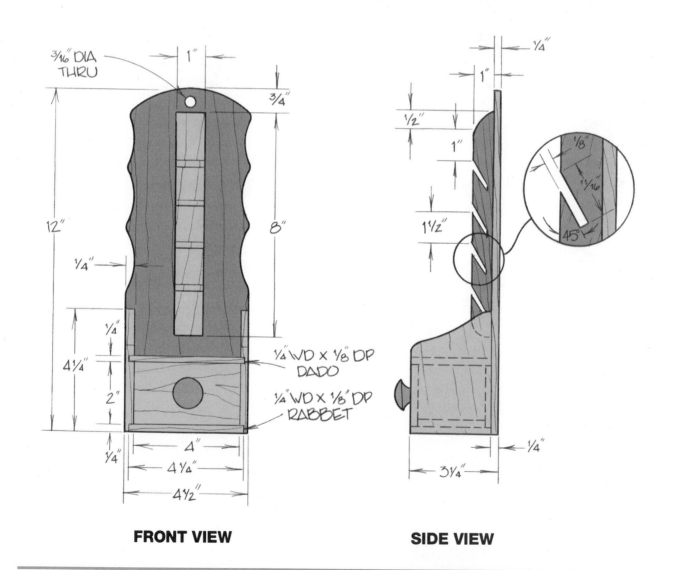

FRONT VIEW   **SIDE VIEW**

4 **Cut the shaped parts.** Enlarge the *Back Pattern* and *Side Pattern,* then transfer them to the stock. Cut the shapes with a band saw or scroll saw, and drill a ³⁄₁₆″-diameter hole through the back to hang the completed holder. Round the ends of the bracket, as shown in the *Side View*. Sand all the sawed edges.

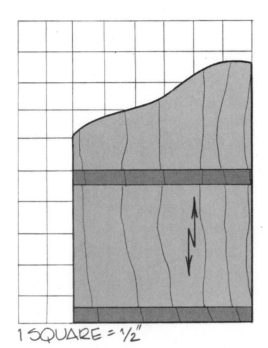

1 SQUARE = ½″

SIDE PATTERN

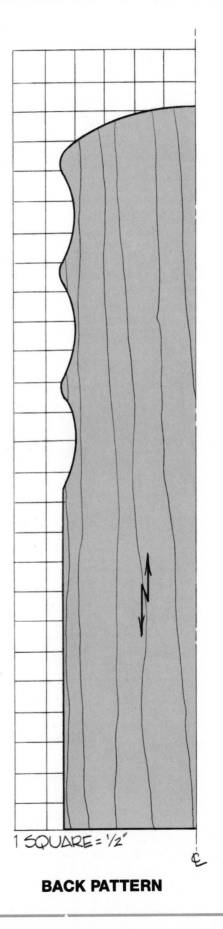

1 SQUARE = ½″

BACK PATTERN

5

Assemble the holder. Finish sand all parts, being careful not to round the edges. Glue the top and bottom to the sides, and reinforce the joints with brads. Set the heads of the brads. Then glue the assembly to the back, flush with the bottom edge.

Center the bracket on the back, ¾″ from the top. Secure it with glue and screws, driving the screws through the back and into the bracket. The screws must not show from the front.

TRY THIS! If you don't want to see the heads of the brads, use a blind nail plane to hide them. Wherever you need to drive a brad, cut a tiny sliver or curl of wood with the plane. Hammer the brad through the hollow left by the curl, then glue the curl back in place. After the glue dries, sand the area clean. The brad will be hidden just under the surface of the wood.

6

Assemble and fit the drawer. Glue the drawer parts together, and reinforce the joints with brads. Secure a drawer pull to the front of the drawer, then test fit the drawer to the holder assembly. Sand a little wood from the sides and top of the drawer until it pulls in and out without binding.

Note: Remember that wood swells in the summer and contracts in the winter. If you build this project in the winter, fit the drawer so there's a ¹⁄₁₆″ gap at the top and sides. If you build it in the summer, you can fit the drawer so it isn't quite so loose.

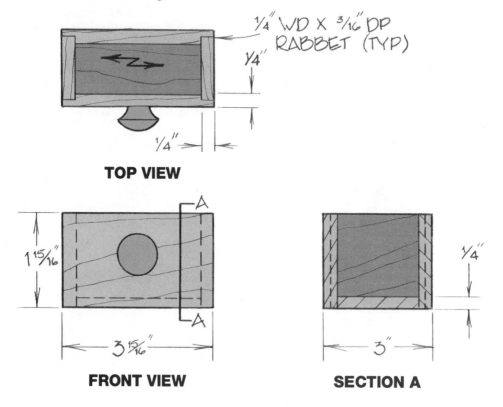

TOP VIEW

FRONT VIEW

SECTION A

7

Finish the holder. Remove the drawer from the assembly, and detach the drawer pull. Do any necessary touch-up sanding, then apply a finish to all wood surfaces, inside and out. Be careful not to let any finish build up inside the mail-holding slots. Wax and buff the finish when it dries, and replace the drawer and drawer pull.

Adjustable Bookcase and Cabinet

Of the many ways to make adjustable shelves, the most common is to rest the shelves on movable supports — clips, brackets, pins, or ledgers. The bookcase shown uses a variation of this idea. The vertical strips inside the case are notched to hold horizontal cleats, and each shelf is suspended between two cleats.

Although this adjustable shelving system is unusual today, it was often used late in the nineteenth century. Most business furniture was made by hand, and individual pieces occasionally were designed for important employees. This bookcase was made especially for a federal government official, the Director General of Railroads.

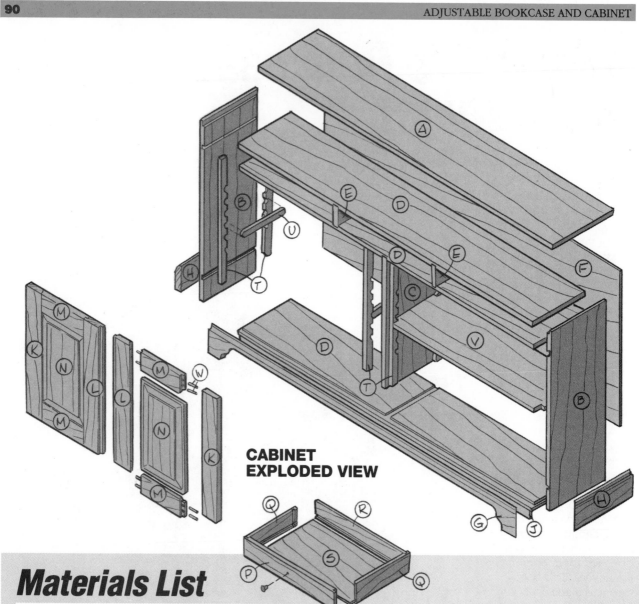

CABINET EXPLODED VIEW

Materials List

FINISHED DIMENSIONS

PARTS

Cabinet

A.	Cabinet top	³⁄₄″ x 12″ x 61½″
B.	Cabinet sides (2)	³⁄₄″ x 11¼″ x 30¼″
C.	Cabinet partition	³⁄₄″ x 11″ x 21¾″
D.	Drawer supports/bottom (3)	³⁄₄″ x 11″ x 59¼″
E.	Drawer dividers (2)	³⁄₄″ x 3¾″ x 11″
F.	Cabinet back	¼″ x 26¼″ x 59¼″
G.	Front cabinet molding	³⁄₄″ x 4³⁄₁₆″ x 61½″

H.	Side cabinet moldings (2)	³⁄₄″ x 4³⁄₁₆″ x 12″
J.	Front cleat	³⁄₄″ x 1½″ x 58½″
K.	Narrow cabinet door stiles (4)	³⁄₄″ x 3″ x 21¼″
L.	Wide cabinet door stiles (4)	³⁄₄″ x 3⁵⁄₁₆″ x 21¼″
M.	Cabinet door rails (8)	³⁄₄″ x 3″ x 8¹⁵⁄₁₆″
N.	Cabinet door panels (4)	³⁄₄″ x 15⁷⁄₈″ x 8¹¹⁄₁₆″
P.	Drawer fronts (3)	³⁄₄″ x 2⁷⁄₈″ x 18⁷⁄₈″

Q.	Drawer sides (6)	½″ x 2⁷⁄₈″ x 10⁵⁄₈″
R.	Drawer backs (3)	½″ x 2⁷⁄₈″ x 18³⁄₈″
S.	Drawer bottoms (3)	¼″ x 10″ x 18³⁄₈″
T.	Cabinet standards (8)	³⁄₄″ x 1″ x 21″
U.	Shelf supports (4–8)	³⁄₄″ x 1″ x 9¼″
V.	Shelves (2–4)	³⁄₄″ x 10⅛″ x 28¾″
W.	Dowels (32)	³⁄₈″ dia. x 3″

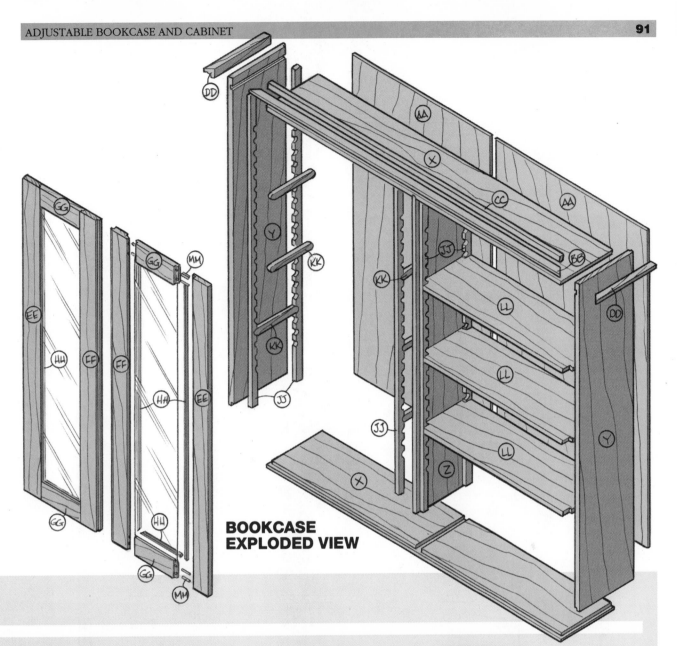

**BOOKCASE
EXPLODED VIEW**

Bookcase

X. Bookcase top/
bottom (2) ³⁄₄″ x 11″ x 59¼″

Y. Bookcase
sides (2) ³⁄₄″ x 11¼″ x 54″

Z. Bookcase
partition ³⁄₄″ x 11″ x 52¼″

AA. Bookcase back
sections (2) ¼″ x 53″ x 29⅝″

BB. Bookcase front
molding 1⅛″ x 1⅜″ x 62¼″

CC. Front molding
support ³⁄₄″ x 1″ x 58½″

DD. Bookcase side
moldings (2) 1⅛″ x 1⅜″ x 12⅜″

EE. Narrow bookcase
door stiles (4) ³⁄₄″ x 3″ x 51¾″

FF. Wide bookcase
door stiles (4) ³⁄₄″ x 3⁵⁄₁₆″ x 51¾″

GG. Bookcase door
rails (8) ³⁄₄″ x 3″ x 8¹⁵⁄₁₆″

HH. Glazing
strips (total) ⅜″ x ½″ x 360″

JJ. Bookcase
standards (8) ³⁄₄″ x 1″ x 51½″

KK. Shelf
supports (8–12) ³⁄₄″ x 1″ x 9¼″

LL. Shelves (4–6) ³⁄₄″ x 10⅛″ x 28¾″

MM. Dowels (32) ⅜″ dia. x 3″

HARDWARE

#10 x 1¼″ Flathead wood screws (72–96)
6d Finishing nails (½ lb.)
1″ Wire brads (1 box)
1½″ x 3″ Butt hinges and mounting
 screws (16)
Cabinet door catches (8)
Key pulls (4)
Drawer pulls (6)
Turnbuttons and screws (16)
⅛″ x 8¹³⁄₁₆″ x 46⅜″ Glass panes (4)

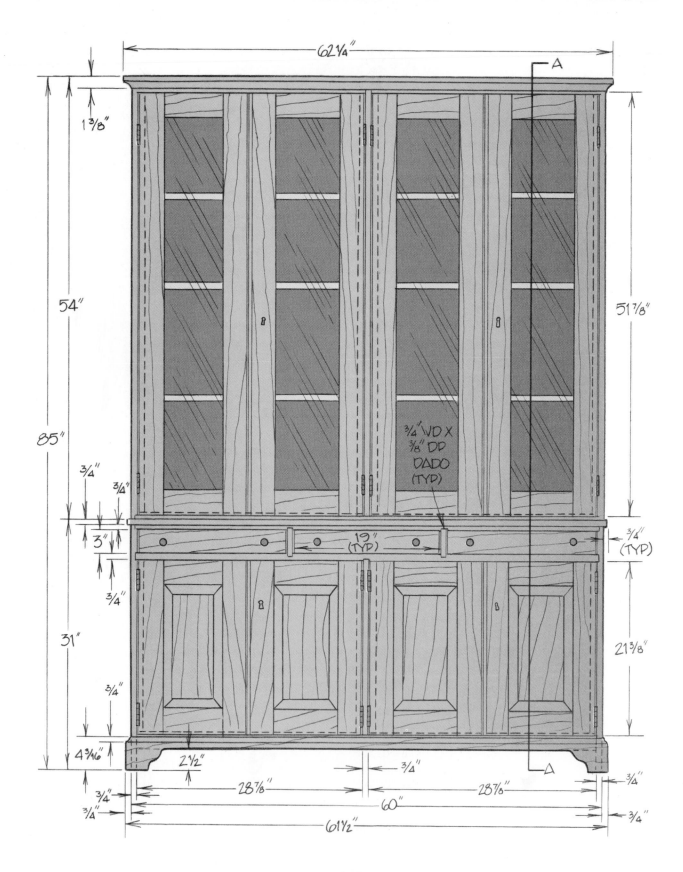

FRONT VIEW

1 **Select the stock.** To make this project, you'll need approximately 120 board feet of 4/4 (four quarters) stock, 5 board feet of 8/4 (eight-quarters) stock, and two 4' x 8' sheets of 1/4" plywood. The cabinet base will require 56 board feet of the 4/4 stock and one sheet of plywood, while the bookcase top will use the remainder. The project shown was built from white pine, although almost any cabinet-grade lumber will do. In the Victorian era, many of these bookcases were built from oak, walnut, or cherry. Remember to purchase plywood with a matching veneer.

Resaw 10 board feet of the 4/4 stock in half, then plane it to 3/8" thick to make the glazing strips. Plane another 4 board feet to 1/2" thick to make the drawer sides and back, and the rest to 3/4" thick. Plane the 8/4 stock to 1 3/8" to make the bookcase molding. You'll find it's easier to build this project in three parts — cabinet base first, then the bookcase, then the doors and drawers. Cut the parts as you build. If you attempt to build everything at once, you may find yourself swimming in wooden parts.

Making the Cabinet

2 **Cut the parts for the cabinet.** Cut all the parts for the cabinet as listed in the Materials List, except the moldings, the door parts, and the drawer parts. Rip the base moldings to width, but don't cut them to length yet. Wait until you've finished the cabinet to cut and miter them.

3 **Cut the dadoes and rabbets in the cabinet parts.** Most of the cabinet parts are joined with simple dadoes and rabbets, reinforced with screws or nails. Make these joints with a dado cutter or a router and a straight bit. Here's a list of the joints and parts to cut:
- 3/4"-wide, 3/8"-deep rabbet in the front edge of the bottom, as shown in the *Cabinet Bottom Layout*
- 3/4"-wide, 3/8"-deep rabbet in the top edge of the sides, as shown in the *Cabinet Side Layout*
- 3/4"-wide, 3/8"-deep dadoes in the sides and bottom,

as shown in the *Cabinet Side Layout* and *Cabinet Bottom Layout*
- 3/4"-wide, 3/8"-deep dadoes in the drawer supports, as shown in the *Drawer Support Layout*
- 3/8"-wide, 3/8"-deep rabbets in each end of the bottom, as shown in the *Cabinet Bottom Layout*
- 3/8"-wide, 3/8"-deep dadoes in the sides, as shown in the *Cabinet Side Layout*
- 1/4"-wide, 3/8"-deep rabbets in the back edges of the sides, as shown in the *Cabinet Side Layout*

CABINET BOTTOM LAYOUT

4

Assemble the cabinet. Assemble the cabinet without glue to check the fit of the parts. Use band clamps and bar clamps to hold the parts together temporarily. When you're satisfied with the fit, finish sand the parts. Reassemble the cabinet with glue, screws, and brads, putting the parts together in this order.

First, attach the drawer dividers to the upper drawer support. Next, attach the top to the upper support;

make sure the back edge of the support is ¼″ in from the back edge of the top, and that the support is centered side-to-side under the top. Attach the lower drawer support to the dividers, then join this assembly to the cabinet sides and bottom. Slide the cabinet partition in place between the lower drawer support and bottom. Check that the cabinet assembly is square, then attach the back with wire brads. *Don't* glue the back in place. Finally, attach the front cleat to the bottom.

TRY THIS! To reinforce the dado joints that join the sides to the drawer supports and bottom, install #10 x 1¼″ flathead wood screws *inside* the cabinet. This will hide them, so they won't show on the outside. Drive the screws at a steep angle, through the horizontal parts and into the sides.

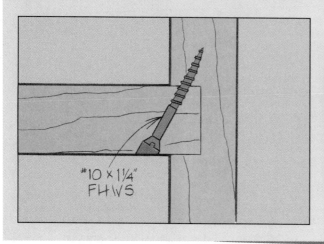

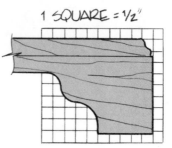

5

Make and install the base molding. With a router or a shaper, cut an ogee shape in the top edge of the base molding stock, as shown in the *Bottom Molding Detail*. Carefully measure the assembled cabinet. The measurements will probably have changed

slightly from those shown in the drawings; this is normal for large projects. Cut the base moldings to fit the cabinet, mitering the adjoining ends.

Enlarge the *Base Molding Pattern* and trace it on the front molding stock. Cut the shape with a band saw or saber saw, then sand the sawed edges.

Attach the moldings to the sides and to the front cleat with glue and screws. Drive the screws from *inside* the cabinet, through the sides or front cleat, and into the moldings. This will hide the screws.

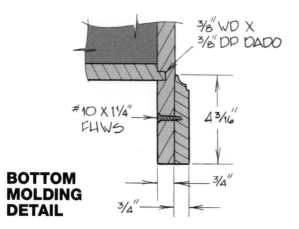

BOTTOM MOLDING DETAIL

⅜″ WD X ⅜″ DP DADO

#10 X 1¼″ FHWS

4³⁄₁₆″

¾″

¾″

1 SQUARE = ½″

BASE MOLDING PATTERN

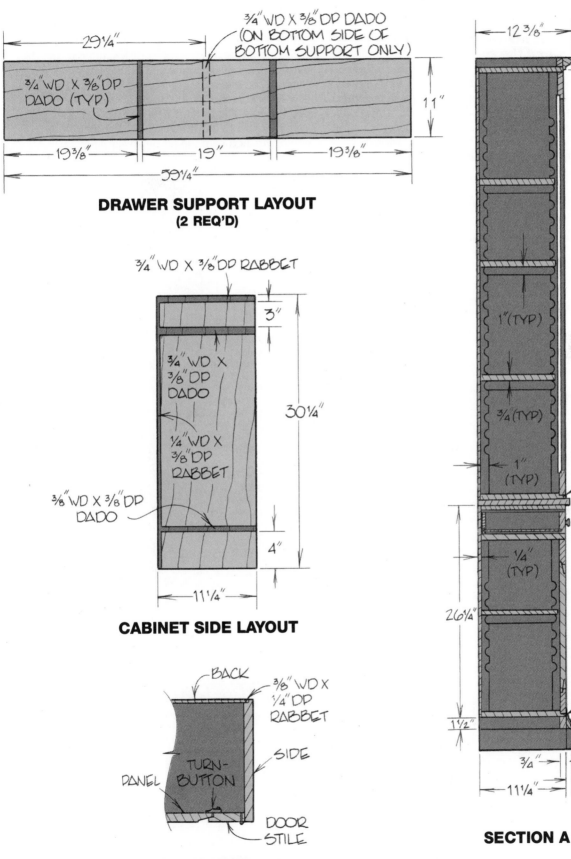

DRAWER SUPPORT LAYOUT
(2 REQ'D)

CABINET SIDE LAYOUT

CABINET JOINERY DETAIL

SECTION A

6

Make the standards, the shelf supports, and the shelves. Measure and mark the position of the scallops on the standards, as shown in the *Cabinet Standard Layout*. Form scallops in two standards at a time. First, clamp two standards together, inside edge to inside edge. Then drill 1″-diameter holes through the seam between the boards, wherever you've marked a scallop. (See Figure 1.) When you take the boards apart, the holes will form the scallops.

Attach the standards to the cabinet sides and partition with glue and screws. Using a disk sander, round the ends of the shelf supports and fit them to the scalloped standards. (See Figure 2.)

Mark the notches on the shelves, as shown in the *Shelf Layout*. Cut the notches with a hand saw or saber saw, then test fit the shelves in the cabinet. When you're satisfied with the fit, finish sand the shelves.

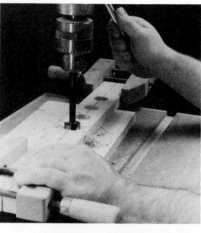

1/To make the scallops in the standards, clamp two standards together edge to edge. Drill 1″-diameter holes through the seam between the boards, wherever you want a scallop. If the standards spread apart as you drill them, hold them with a hand screw near the scallop you're about to drill.

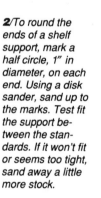

CABINET STANDARD LAYOUT

2/To round the ends of a shelf support, mark a half circle, 1″ in diameter, on each end. Using a disk sander, sand up to the marks. Test fit the support between the standards. If it won't fit or seems too tight, sand away a little more stock.

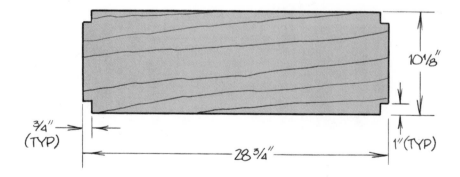

SHELF LAYOUT

Making the Bookcase

7 ***Cut the parts for the bookcase.*** The bookcase is made with almost the same procedure you used for the cabinet. Cut all the parts for the book-case, as specified in the Materials List, except the moldings and the door parts. Rip the base moldings to the proper width, but don't cut them to length.

8 ***Cut the dadoes and rabbets in the bookcase parts.*** Like the cabinet, the book-case is largely assembled with dadoes and rabbets. Here's a list of the joints and parts to cut:
- ¾"-wide, ⅜"-deep rabbet in the front edge of the bottom, as shown in the *Bookcase Bottom Layout*
- ¾"-wide, ⅜"-deep dadoes in the bookcase top and bottom, as shown in the *Bookcase Top Layout* and *Bookcase Bottom Layout*
- ¾"-wide, ⅜"-deep dadoes in the sides, as shown in the *Bookcase Side Layout*
- ⅜"-wide, ⅜"-deep rabbets in each end of the bottom, as shown in the *Bookcase Bottom Layout*
- ⅜"-wide, ⅜"-rabbets in the sides, as shown in the *Bookcase Side Layout*
- ¼"-wide, ⅜"-deep rabbets in the back edges of the sides, as shown in the *Bookcase Side Layout*

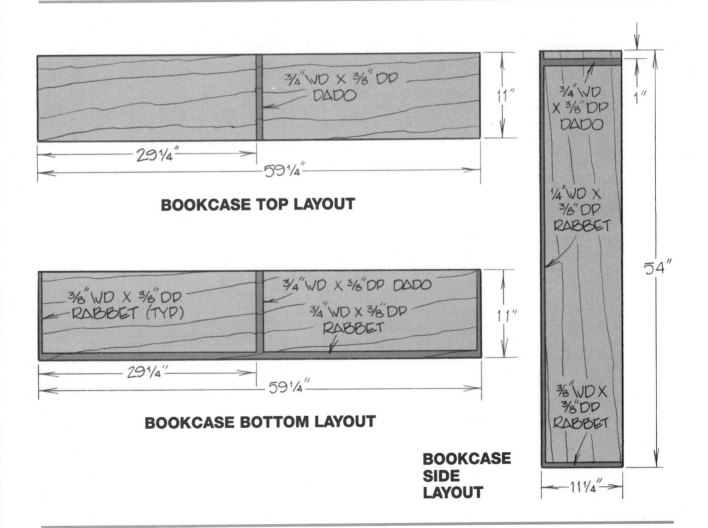

BOOKCASE TOP LAYOUT

BOOKCASE BOTTOM LAYOUT

BOOKCASE SIDE LAYOUT

Assemble the bookcase. Assemble the bookcase without glue to check the fit of the parts. As you did before, use band clamps and bar clamps to hold the parts together temporarily. When you're satisfied with the fit, finish sand the parts. Reassemble the bookcase with glue, screws, and brads, putting the parts together in this order.

First, attach the sides to the top and bottom. Slide the bookcase partition in place between the top and bottom. Check that the bookcase assembly is square, then attach the back panels with wire brads. As before, *don't* glue the back pieces in place. Finally, attach the front molding support to the top.

10 Make and install the top molding. With a router or a shaper, cut a cove in the front face of the top molding stock, as shown in the *Top Molding Detail*. Carefully check the assembled bookcase to see if the measurements have changed. Cut the base moldings to fit the cabinet, mitering the adjoining ends.

Attach the moldings to the sides and to the front support with glue and screws. As before, drive the screws from *inside* the bookcase so they won't be seen.

TOP MOLDING DETAIL

11 Make the standards, the shelf supports, and the shelves. Measure and mark the position of the scallops on the standards, as shown in the *Bookcase Standard Layout*. Clamp two standards together and drill 1″-diameter holes to create the scallops, as you did when making the cabinet.

Attach the standards to the bookcase sides and partition with glue and screws. Using a disk sander, round the ends of the shelf supports and fit them to the scallops in the standards. Mark the notches on the shelves, cut them, and fit them in the bookcase. Finish sand these shelves.

With a helper, place the completed bookcase on top of the cabinet. Drive several screws through the bookcase bottom into the cabinet top to hold the two assemblies together. Do *not* glue these pieces, and don't cover the heads of the screws — leave them accessible. When you need to move the cabinet, you'll be able to take the sections apart by loosening the screws.

BOOKCASE STANDARD LAYOUT

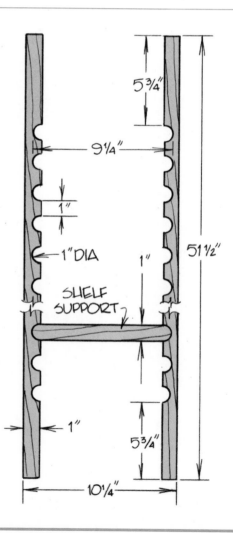

Making the Drawers and Doors

12 **Make and install the drawers.** Cut the drawer parts — fronts, backs, sides, and bottoms. Then cut the joinery:

■ ½"-wide, ⅜"-deep rabbets in the ends of the drawer fronts, as shown in the *Drawer/Top View*
■ ½"-wide, ¼"-deep dadoes in the drawer sides
■ ¼"-wide, ¼"-deep grooves in the inside faces of the fronts, backs, and sides, as shown in the *Drawer/Side View*

Test fit the drawer parts. When you're satisfied that the joints fit properly, drill holes for the pulls in the drawer fronts. Then finish sand all the drawer parts.

To assemble each drawer, first glue the drawer back in the dadoes in the drawer sides. Slide the bottom into the grooves, and glue the sides in the drawer front rab-

bets. *Don't* glue the bottom in place; let it float in the grooves. Reinforce the rabbet joints with finishing nails, driving the nails through the sides and into the front.

Install the drawer pulls in the drawer fronts, then insert each drawer in the cabinet. Test the sliding action. If a drawer binds when pulled out or pushed in, remove some stock from it with a file or a plane.

> **TRY THIS!** Some professional cabinet-makers prefer to make drawers ¹⁄₁₆"–⅛" oversize, then sand them down to fit the cabinet precisely.

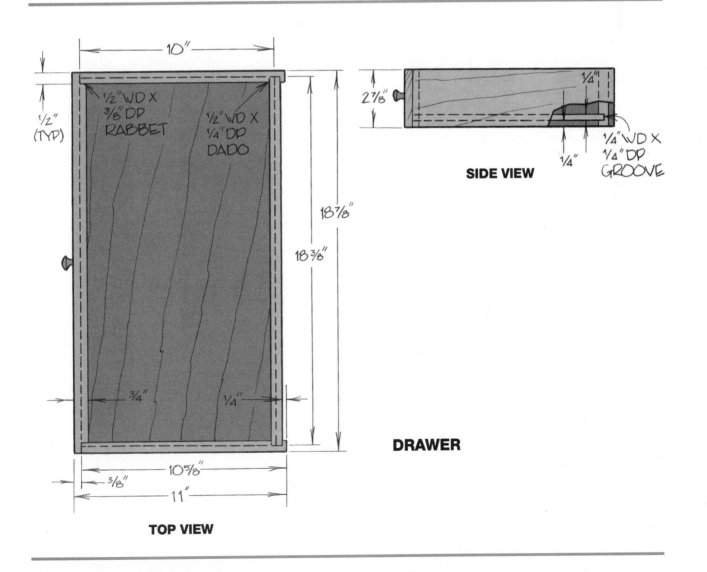

SIDE VIEW

DRAWER

TOP VIEW

13 **Assemble the door frames.** Cut the door parts from the straightest, clearest stock you can find. *This is extremely important!* If a rail or a stile is bowed or warped even slightly, the door may not fit the cabinet or the bookcase properly. It will look out of kilter, no matter how carefully you assemble and hang it.

Note: Do not cut the glazing strips yet.

Carefully mark which parts belong to which frame, and which ends are joined to which edges. Also mark the inside and outside face of each part. This will help you identify the parts you need as you cut them, and prevent you from getting them mixed up.

Sort out the eight wide stiles: When the doors are closed, these stiles lap each other, as shown in the *Door Overlap Detail*. To make these laps, cut matching 5/16"-wide, 3/8"-deep rabbets in the adjoining outside edges — one rabbet in the inside surface, and the other in the outside. With a molder or a router, cut a bead on the *outside* surface of the four wide stiles with the *inside* rabbets.

Assemble the door frames with *doweled* rabbet joints. To make each door frame, lay out the rails and stiles on your workbench. Push the parts together, so they touch where you want to join them. Mark the location of each dowel, drawing a line across the joint and marking both the rail and the stile. You should put two dowels in each joint.

Using a doweling jig to guide the drill, make 3/8"-diameter, 1 11/16"-deep dowel holes in the edges of the stiles and the ends of the rails. Dry assemble the frame with dowels to make sure all the dowel holes are properly aligned.

Cut a 3/8"-wide, 3/8"-deep rabbet on the inside edge of the rails and the stiles, then cut the same-size rabbet in the ends of the rails. The rabbets in the ends of the rails should lap the rabbets in the edges of the stiles. (See Figures 3 through 5.)

Finish sand the rails and stiles, then assemble the door frames with glue and dowels. As you clamp the parts together, make certain that the frames are square.

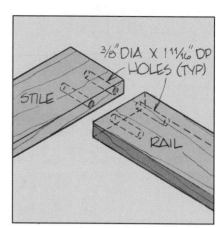

3/To make doweled rabbet joints, start by drilling the dowel holes. Use a doweling jig to guide the drill, and make enough holes to install at least two dowels in each joint.

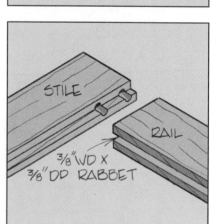

*4/After you drill and test fit the dowel joints, cut rabbets in the edge of the rails **and** stiles. These rabbets will eventually hold the door panels or glass.*

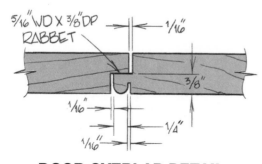

*5/Finally, cut matching rabbets in the ends of the rails **only**. These end rabbets must lap the edge rabbets so the surfaces of the assembled rails and stiles are flush.*

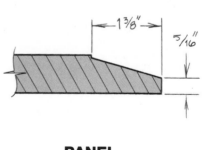

DOOR OVERLAP DETAIL

PANEL SECTION

14 Make and install the raised panels and glazing strips.

While the glue is drying on the frames, make the wooden door panels. The panels in the bookcase doors are clear glass, but those in the cabinet doors are solid wood *raised* panels, as shown in the *Panel Section*. There are several ways to raise these panels, using a router, shaper, or table saw. The most common method is to bevel cut the edges and ends of each panel with a table saw. Use the rip fence and a tall fence extension to guide the stock, and adjust the depth of cut so the saw leaves a ⅛″ step between the flat and the beveled areas of the panel. (See Figure 6.)

Install the wood panels in the cabinet door frames, securing them with turnbuttons. This will allow the panels to expand and contract with changes in the weather. Do *not* install the glass panels in the bookcase door frames until after you've applied the finish.

You can, however, cut the glazing strips that hold the glass in the doors. With a router and radius edge-cutter bit, round over one edge of the ⅜″-thick glazing stock. Then rip a ½″-wide strip from the board. (See Figure 7.) Repeat until you have made all the glazing strips needed.

6/When cutting a raised panel with a table saw, adjust the angle of the blade to approximately 15°. Saw the bevels with a hollow-ground planer blade so the cuts will be smooth and clean.

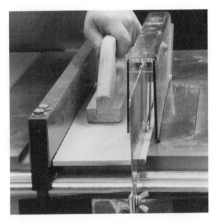

7/When making narrow moldings such as the glazing strips, shape the edge of a wide board, **then** rip the narrow stock from the board. If you try to shape narrow stock, it may splinter or break up while you cut it.

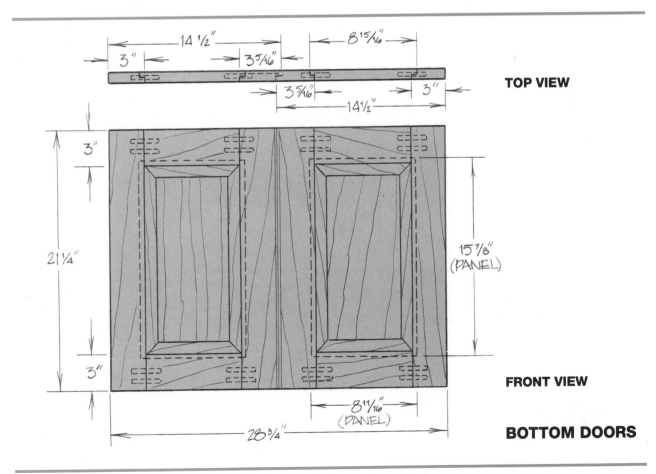

TOP VIEW

14 ½″   8¹⁵⁄₁₆″
3″   3⁵⁄₁₆″
3⁵⁄₁₆″   3″
14½″

3″

21¼″

15⅞″ (PANEL)

3″

8¹¹⁄₁₆″ (PANEL)

28¾″

FRONT VIEW

BOTTOM DOORS

15 **Hang the doors.** Cut hinge mortises in the outside edges of the narrow stiles with a sharp chisel. Make matching mortises in the inside surfaces of the bookcase and cabinet sides. Attach the hinges to the doors, then hang the doors in the cabinet. Install door pulls and catches.

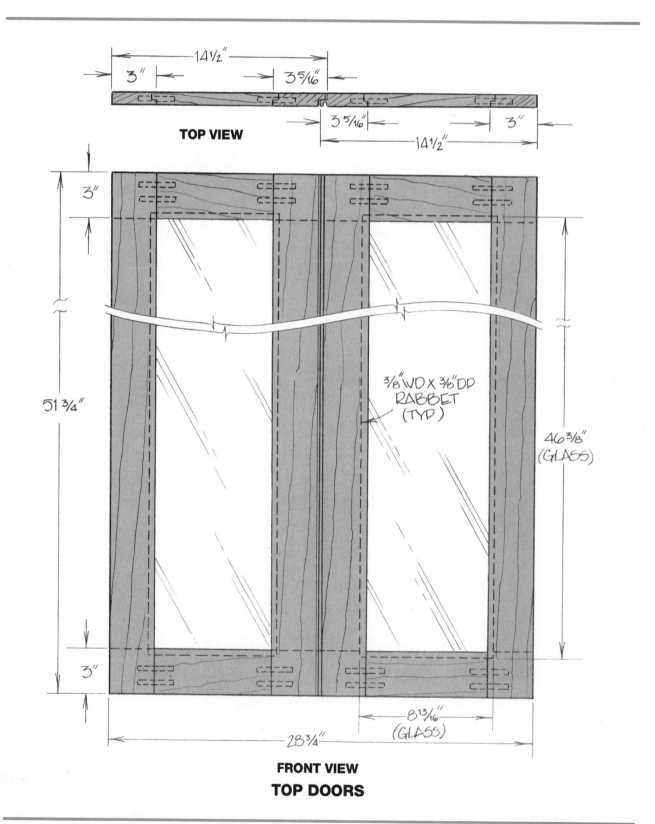

TOP VIEW

3/8"WD X 3/8"DP
RABBET
(TYP)

46 3/8"
(GLASS)

8 13/16"
(GLASS)

FRONT VIEW
TOP DOORS

Finishing Up

16 Finish the bookcase and cabinet.

Remove the shelves, shelf supports, doors, and drawers from the bookcase and the cabinet. Detach all the hardware — pulls, catches, hinges — and set it aside. Loosen the screws that hold the two sections together and take them apart.

Do any necessary touch-up sanding, then apply paint, stain, or another finish. The project shown is stained a cherry color, then coated with varnish. The bookcase and cabinet are painted a flat, dark brown on the inside, which makes the wooden standards and the shelf supports less noticeable.

As you finish the project, remember to include the glazing strip stock, although you haven't yet installed it. Be sure to apply as many coats to the inside of the bookcase and cabinet as you do to the outside. Let the finish dry, then polish it with wax.

17 Assemble the project and install the glass panels.

Assemble all the sections and parts, and install the hardware. To secure the glass panels in the bookcase doors, first cut the glazing strips to fit in the rabbets and miter the adjoining ends. Put the glass in place, then attach the strips to the rails and stiles with wire brads, as shown in the *Glazing Detail*. Set the heads of the brads below the surface of the wood. Finally, touch up the finish wherever you may have accidentally scratched the wood or hit it with a hammer.

GLAZING DETAIL

Variations

If you don't wish to make wooden standards and supports, you can purchase hardware to support adjustable shelving. Here are three common types: *Pin supports (A)* rest in holes drilled into the sides of the case, and the shelf rests on the protruding pin.

Wire supports (B) also rest in holes, and the ends of the shelves are slotted to fit over the wires. *Pilasters (C)* mount in grooves cut in the sides, and the shelves rest on *clips* that attach to the pilasters.

Common Shelving Support Hardware/These shelving supports are available from most hardware and building supply stores, as well as from mail-order woodworking suppliers. They are (A) pin supports, (B) wire supports, and (C) pilasters and clips.

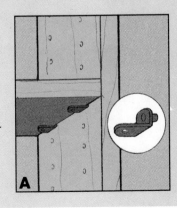

A

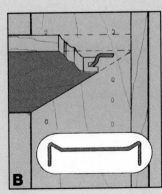

B

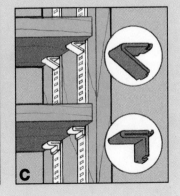

C

Drafting Table

Desks and tables are designed to help you deal with letters, bills, and other *small* documents. However, they aren't quite as useful for working with larger pieces of paper. If you lay an ordinary set of building plans flat on a desk, the bottom portion of the plans will be much closer to you than the top portion. You must constantly focus and refocus your eyes as you look over the drawings. Before long, your eyes are apt to become extremely tired. Your neck may also feel the strain.

A drafting table solves this problem. The work surface tilts so that everything is in the same focal plane — approximately the same distance from your eyes. You may still have to refocus as you look from one end of the plans to the other, but not nearly so much. Your eyes (and your neck) do less work, and you can see better longer.

The drafting table shown was built by Virgil Schmidt of Vandalia, Ohio. Virgil designs and builds furniture in his cabinetmaking shop, Lifestyles. "When I began woodworking," says Virgil, "just about the first thing I built was this drafting table. I found it almost impossible to lay out my projects without it."

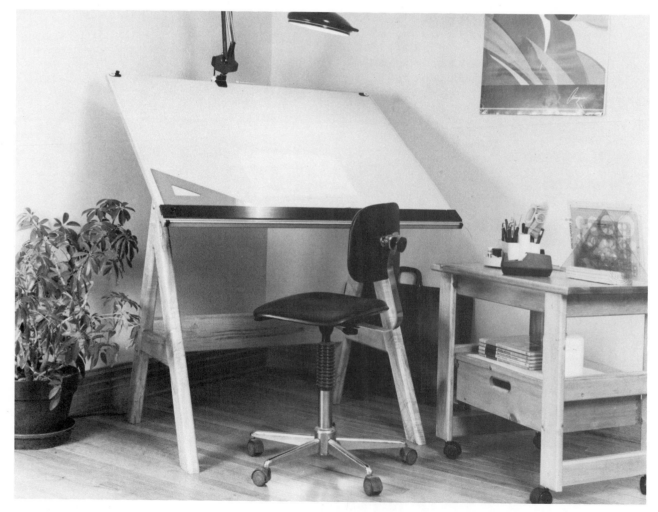

EXPLODED VIEW

Materials List

FINISHED DIMENSIONS

PARTS

A.	Tabletop	³⁄₄″ x 32″ x 44″
B.	Front/back edging (2)	³⁄₄″ x 2″ x 44″
C.	Side edging (2)	³⁄₄″ x 2″ x 36″
D.	Pivots (2)	1½″ x 4¼″ x 32″
E.	Stiffeners (2)	1½″ x 2″ x 41″
F.	Trunnions (2)	³⁄₄″ x 9″ x 14″
G.	Long trunnion mounts (2)	³⁄₄″ x ³⁄₄″ x 14″
H.	Short trunnion mounts (4)	³⁄₄″ x 1½″ x 3½″
J.	Clamps (2)	³⁄₄″ x 1½″ x 4″
K.	Front legs (2)	1½″ x 3″ x 36³⁄₈″
L.	Back legs (2)	1½″ x 3″ x 45″
M.	Braces (2)	1½″ x 3″ x 13³⁄₈″
N.	Stretcher	1½″ x 6″ x 38″
P.	Splines (total)	¼″ x 1″ x 168″

HARDWARE

#10 x 2″ Flathead wood screws (4)
#10 x 1¼″ Flathead wood screws (24–30)
#20 Biscuits (12)
³⁄₈″ x 4″ Lag screws (4)
³⁄₈″ x 3½″ Carriage bolts (6)
³⁄₈″ Flat washers (6)
³⁄₈″ Wing nuts (6)
36″ x 48″ Vinyl pad (optional)

1

Change the size of the table to suit your needs. The table shown was built for an adult who uses it in his profession. Children may find it too large to work at comfortably. Some adults may find it too big for their purposes, too. Not everyone has a need — or the room — for a full-size drafting table.

To build a scaled-down version, keep the trunnion parts the same size, but reduce the length *and* width of the tabletop, and the length *only* of the remaining parts.

Shrink them all by the *same proportion*. For example, if you reduce all these dimensions to ¾ of full size, the drafting table will be 36″ wide, 27″ deep, and 32¾″ high. This occupies much less space, but the table is still comfortable for adults and teenagers. If you're building it for someone who's 9–13 years of age, reduce the dimensions to ⅔. If the person is younger still, reduce them to ½.

2

Select the stock and cut the parts. To build this drafting table as shown, you'll need approximately 5 board feet of 4/4 (four-quarters) stock, 20 board feet of 8/4 (eight-quarters) stock, some scraps of ¼″ plywood, and one half of a 4′ x 8′ sheet of *Baltic birch* ¾″ plywood. Make the edging and clamps from the 4/4 stock, and the legs, braces, stretcher, and stiffeners from the 8/4 stock. Use the ¼″ plywood scraps for splines, and cut the tabletop and trunnions from the Baltic birch plywood.

Baltic birch is a special laminated-veneer plywood that is extremely strong and stable. It helps ensure that the table will remain as flat as possible, and the trunnions will be as strong as possible. This type of plywood isn't always readily available, but most lumberyards will special-order it for you. You may also be able to order it through a local cabinetmaker.

The type of wood you use for the other parts is not as critical. You can use almost any hardwood or softwood for the legs, braces, and stretcher. On the drafting table

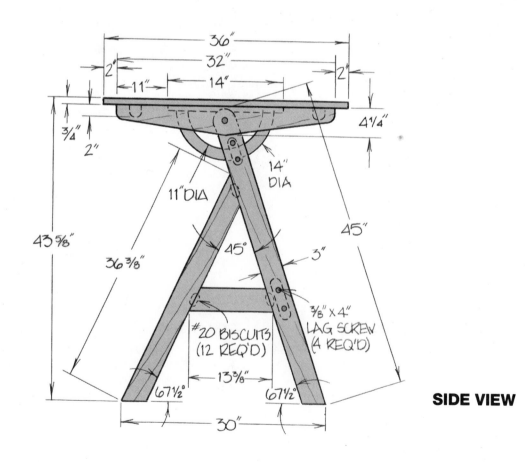

SIDE VIEW

shown, these parts are made from sugar maple. Choose a harder wood, such as birch or rock maple, for the edging — this will prevent the edges of the table from getting beat up.

When you have purchased the materials, plane the 8/4 stock to 1½" thick, and the 4/4 stock to just a hair over ¾" thick. Since much of the 4/4 stock will be used for edging, you want to make it slightly thicker than the

plywood. After you glue it in place, you can scrape and sand the solid wood flush with the plywood. If you have to scrape and sand the plywood, you may wear through the veneer.

Cut all the parts to the lengths and widths shown in the Materials List. Miter the ends of the legs and braces as shown in the *Side View*.

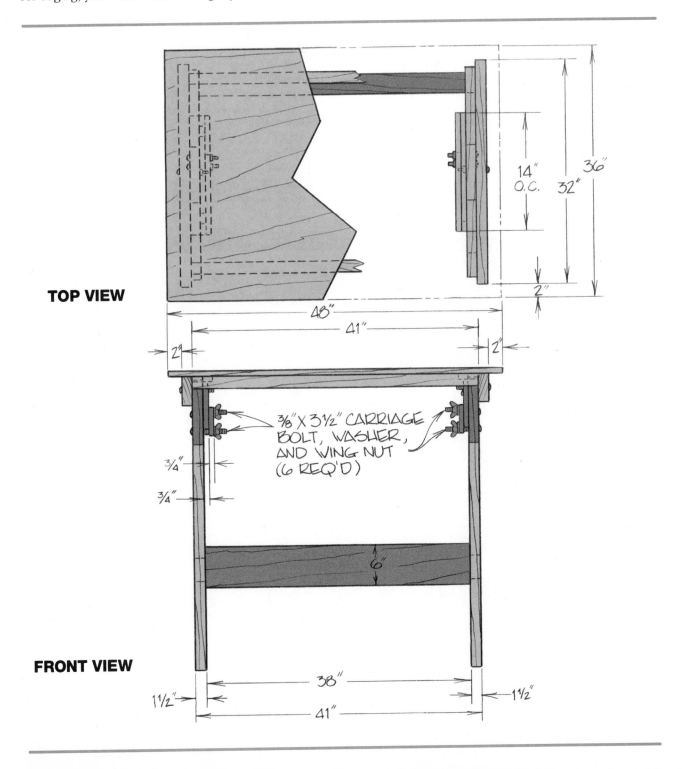

TOP VIEW

FRONT VIEW

3/8" X 3½" CARRIAGE BOLT, WASHER, AND WING NUT (6 REQ'D)

3 Drill holes in the pivots, back legs, and clamps.
The tabletop tilts on two ⅜″ carriage bolts. Several other carriage bolts secure the trunnions, holding the table at the desired angle. Carefully mark the locations of the holes for these bolts. Also mark the round ends of the back legs, but don't cut them yet. First, drill the bolt holes:

- One ⅜″-diameter hole through each pivot, as shown in the *Pivot Layout*
- Two ⅜″-diameter holes through each clamp, as shown in the *Clamp Layout*
- Three ⅜″-diameter holes through each back leg, as shown in the *Back Leg Layout*

4 Cut the shape of the back legs, pivots, clamps, and trunnions.
Lay out the shape of the trunnions and the pivots, as shown in the *Trunnion Layout* and *Pivot Layout*. Using a band saw or a saber saw, cut the shapes — the round ends of the back legs and clamps, the outside shape of the trunnions, and the shape of the pivots. Sand the sawed edges to remove the saw marks.

Drill a hole in the waste inside each trunnion. Insert a saber saw blade in the hole, and saw up to the line that marks the inside shape. Cut along the line and remove the waste. Once again, sand the sawed edges.

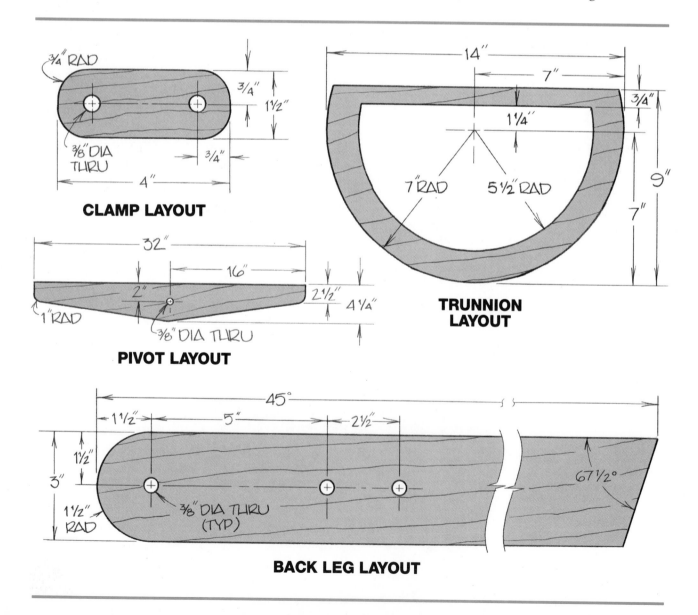

CLAMP LAYOUT

TRUNNION LAYOUT

PIVOT LAYOUT

BACK LEG LAYOUT

5 Round over the stiffeners and stretcher.

With a router and a ½″ quarter-round bit, round over the *bottom* edges of both stiffeners, and *both* edges of the stretcher. Sand the rounded areas to remove any mill marks.

6 Attach the edging to the tabletop.

Using a router and a slotting cutter, make ¼″-wide, ½″-deep grooves in all the inside edges of the edging stock, and the *front* and *back* edges of the tabletop. *Don't* rout the side edges yet. Glue the top and bottom edging to the tabletop with splines. (See Figures 1 and 2.) Wipe away any glue that squeezes out of the spline joints with a wet rag.

When the glue dries, sand the joints clean and flush. In particular, make sure the ends of the top and bottom edging are flush with the sides of the tabletop. Rout the sides, cutting grooves in both the tabletop and the ends of the attached edging. Glue the side edging to the top assembly with splines. Wait for the glue to dry, then sand the joints.

1/Cut spline grooves in the adjoining edges of the tabletop and the edging with a slotting cutter. These cutters are larger than most router bits and, because of this, they generate a good deal more torque. Make sure you have a firm grip on the router as you work. Cut **against** the direction of rotation.

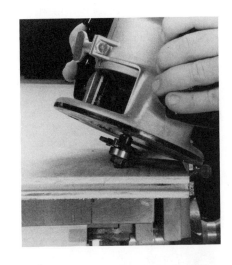

2/Insert splines in the grooves and join the edging and the tabletop. For the strongest possible spline joints, make the splines from plywood. If you make them from solid wood, make sure the wood grain of each spline runs **perpendicular** to the length of the groove.

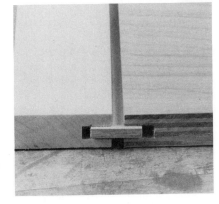

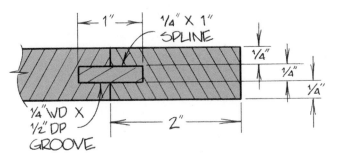

EDGE JOINERY DETAIL

7 Assemble the trunnions and trunnion mounts.

Drill screw pilot holes in both the long and the short trunnion mounts, so you can attach them to the underside of the tabletop. Countersink these holes for the heads of the screws.

Finish sand the mounts and the trunnions. Glue the mounts to the trunnions, as shown in the *Trunnion Assembly Detail*. When the glue dries, sand the upper surface of each assembly clean and flush.

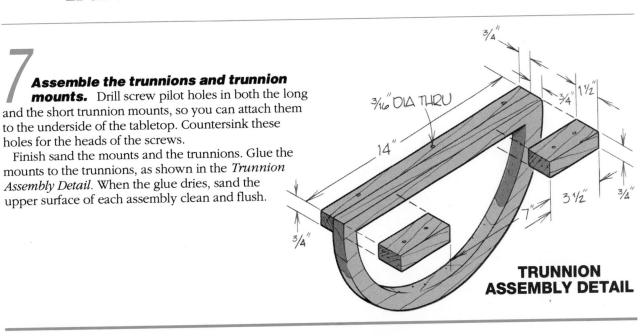

TRUNNION ASSEMBLY DETAIL

8 **Assemble the legs.** Finish sand the back legs, front legs, and braces. Lay out one set of legs and a brace on your workbench. Temporarily nail or clamp a 30″-long scrap to one end of the bench, and position the bottoms of the legs flush against this scrap. Make sure all the joints fit properly, then mark where the parts join each other. Also mark where you will insert the wooden plates or biscuits between the parts.

Using a plate joiner, cut slots for #20 biscuits in the adjoining ends and edges. Use two biscuits per joint, one biscuit about ½″ in from the outside face, and the other ½″ in from the inside face. (See Figure 3.) Apply glue to the slots and the adjoining surfaces. Insert biscuits in the slots, then assemble the back leg, front leg, and brace. Let the glue dry, then sand the joints clean and flush.

Lay out the second set of legs and brace on top of the first, so the two assemblies will be precisely the same. Mark the parts and join them with biscuits, as before.

Round the inside and outside edges of the leg assemblies with a router and a ½″ quarter-round bit. Sand the edges to remove the mill marks.

*3/*To make the leg assemblies as strong as possible, cut two slots in each of the adjoining surfaces, as shown. Use two #20 biscuits in each joint.

9 **Assemble the drafting table.** Finish sand any parts that still need it. Carefully position the pivots and stiffeners on the underside of the tabletop assembly. Mark their positions, then attach them to the table with glue and screws. Drive the screws down through the table and into the solid wood parts. Counterbore *and* countersink these screws, then cover the heads with wooden plugs. Sand the plugs flush with the top surface of the table.

With a helper, attach the leg assemblies to the pivots with carriage bolts, washers, and wing nuts. Hand-tighten the nuts as much as possible, then turn the table assembly right side up and rest it on its legs. Attach the stretcher between the back legs with lag screws and washers. *Don't* glue the stretcher in place or cover the

heads of the lag screws. This will allow you to knock down the drafting table whenever you want to move or store it.

Attach the trunnion assemblies to the underside of the tabletop with just *two* screws each. Insert the carriage bolts that hold the clamps through the back legs, then tilt the tabletop up and down. As you do so, watch the trunnions. Make sure they don't rub on the carriage bolts. As the table pivots, the trunnions should remain the same distance from each bolt. If they don't, loosen the screws and reposition the offending trunnion assembly. When the trunnions are positioned properly, install all the screws. Also install the clamps, along with the remaining washers and wing nuts.

10 **Finish the drafting table.** Disassemble the drafting table, separating the top assembly, leg assemblies, clamps, and stretcher. Remove all the carriage bolts, lag screws, and other hardware and set it aside. Apply a durable finish, such as spar varnish or

polyurethane, to all the wooden surfaces. Let the finish dry, rub it out, and apply a coat of wax.

Reassemble the table. If you wish, cover the top surface of the drafting table with a vinyl pad. These pads are available from most artist's and drafting supply stores.

Schoolmarm's Desk

The schoolmarm's desk — sometimes called a lectern desk — evolved to meet the needs of the rural nineteenth-century schoolteacher. Alone in a one-room schoolhouse, the teacher was often the principal, superintendent, and the school board all rolled into one. As such, her responsibilities included not only teaching and lecturing, but administrative chores as well.

A schoolmarm's desk was both a lectern and an office desk. It started out as a simple writing table with a separate stand to hold books and notes. Before long, the stand was built right into the desk. Later refinements included small drawers and pigeonholes to hold writing materials and help organize correspondence.

Larry Callahan, of West Milton, Ohio, modified the traditional design when he built this desk for use in his home office. He adjusted the height and size of the lectern to make it a comfortable writing surface, and redesigned the pigeonholes to hold standard-size correspondence and bills. The desk retains its quaint nineteenth-century charm, but serves twentieth-century needs.

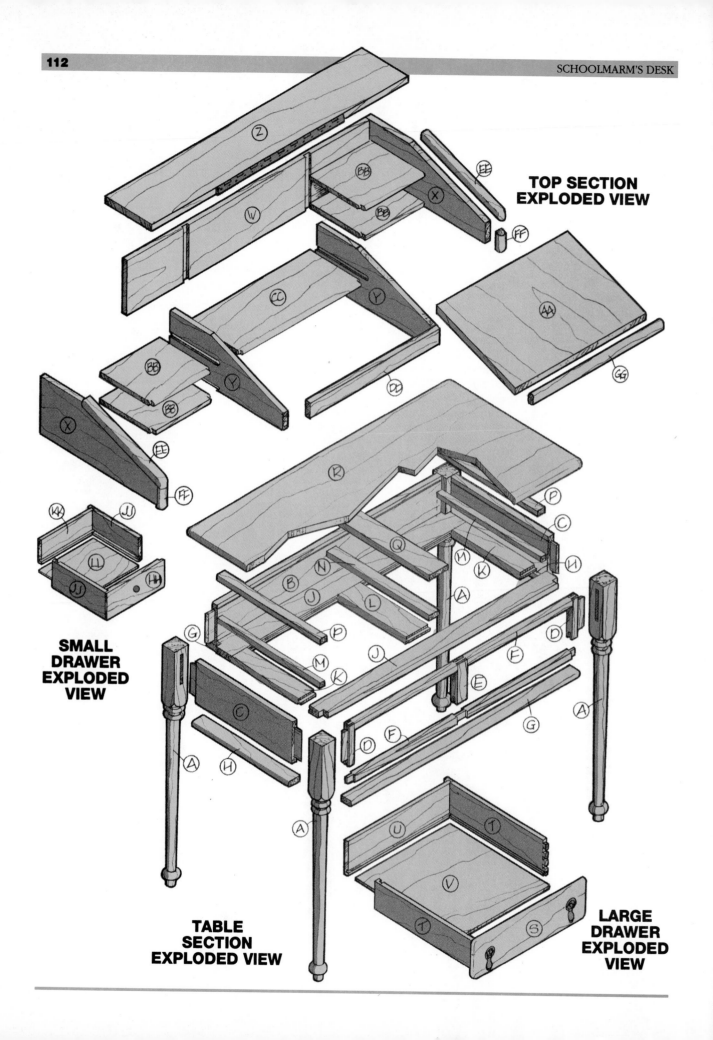

TOP SECTION EXPLODED VIEW

SMALL DRAWER EXPLODED VIEW

TABLE SECTION EXPLODED VIEW

LARGE DRAWER EXPLODED VIEW

Materials List

FINISHED DIMENSIONS

PARTS

Table Section

A.	Legs (4)	2″ x 2″ x 29″
B.	Back apron	3/4″ x 5 1/4″ x 36 1/4″
C.	Side aprons (2)	3/4″ x 5 1/4″ x 16 3/4″
D.	End stiles (2)	3/4″ x 1 7/8″ x 5 1/4″
E.	Middle stile	3/4″ x 1 1/2″ x 5 1/4″
F.	Rails (2)	3/4″ x 3/4″ x 34″
G.	Front/back apron moldings (2)	3/4″ x 1 1/2″ x 34″
H.	Side apron moldings (2)	3/4″ x 1 1/2″ x 14 1/2″
J.	Web frame rails (2)	3/4″ x 2 3/8″ x 35 1/4″
K.	Web frame end stiles (2)	3/4″ x 2 3/8″ x 11 3/4″
L.	Web frame middle stile	3/4″ x 3 1/2″ x 11 3/4″
M.	End drawer guides (2)	3/4″ x 3/4″ x 15 3/4″
N.	Middle drawer guide	3/4″ x 1 1/2″ x 15 3/4″
P.	End kickers (2)	3/4″ x 1″ x 15 3/4″
Q.	Middle kicker	3/4″ x 3″ x 15 3/4″
R.	Desk top	3/4″ x 19 1/2″ x 39″
S.	Large drawer fronts (2)	3/4″ x 4 3/8″ x 16 1/8″
T.	Large drawer sides (4)	1/2″ x 3 5/8″ x 15 3/8″
U.	Large drawer backs (2)	1/2″ x 3 5/8″ x 14 7/8″
V.	Large drawer bottoms (2)	1/4″ x 14 3/4″ x 14 7/8″

Top Section

W.	Back	3/4″ x 7″ x 36″
X.	Sides (2)	3/4″ x 7″ x 17 3/4″
Y.	Dividers (2)	3/4″ x 7″ x 16 5/8″
Z.	Top	3/4″ x 7″ x 38″
AA.	Lid	3/4″ x 12 1/8″ x 19″
BB.	End shelves (4)	1/2″ x 9″ x 8″
CC.	Middle shelf	1/2″ x 8″ x 18″
DD.	Front	3/4″ x 3 3/16″ x 19″
EE.	Upper side moldings (2)	3/4″ x 1″ x 13″
FF.	Front side moldings (2)	3/4″ x 1″ x 2 7/8″
GG.	Lid molding	3/4″ x 1 1/8″ x 19″
HH.	Small drawer fronts (2)	3/4″ x 3 3/8″ x 8 3/8″
JJ.	Small drawer sides (4)	1/2″ x 3 3/8″ x 7 7/8″
KK.	Small drawer backs (2)	1/2″ x 3 3/8″ x 7 7/8″
LL.	Small drawer bottoms (2)	1/4″ x 7″ x 7 7/8″

HARDWARE

1 1/2″ x 19″ Brass piano hinge and mounting screws
Pendant drawer pulls (4)
Small drawer pulls (2)
Tabletop clips (8–12)
#6 x 5/8″ Flathead wood screws
#10 x 1 1/4″ Roundhead wood screws (10–12)
#10 Flat washers (10–12)
4d Finishing nails (30–40)

---

1 Select the stock and cut the parts. To make this project you'll need approximately 35 board feet of 4/4 (four-quarters) lumber, one-half of a 4′ x 8′ sheet of cabinet-grade 1/4″ plywood, and four 3″ x 3″ turning squares approximately 30″ long. (**Note:** If 2 1/4″-square or 2 1/2″-square turning stock is available, buy it instead of 3″ square. There will be less waste.) Traditionally, these desks were often built from walnut or cherry, but you can use almost any hardwood. To save some money, you may want to use an inexpensive wood, such as poplar, for the interior parts — web frame assembly, kickers, drawer sides, and drawer backs. The primary (exterior) wood on the desk shown is cherry, and the secondary (interior) wood is birch.

Plane 8 board feet of the 4/4 stock to 1/2″ thick, to make the shelves and drawer parts, and the remainder to 3/4″ thick. Square the turning stock on a jointer, then plane it to 2″ x 2″ to make the leg stock. Glue up wide stock to make the desk top, then cut the parts to the sizes shown in the Materials List, with two exceptions: (1) cut the legs at least 2″ longer than given, so you have some extra stock to mount them on the lathe; and (2) rip the moldings to width, but don't cut them to length yet. As you cut the lid, bevel the upper and lower edges at 70°.

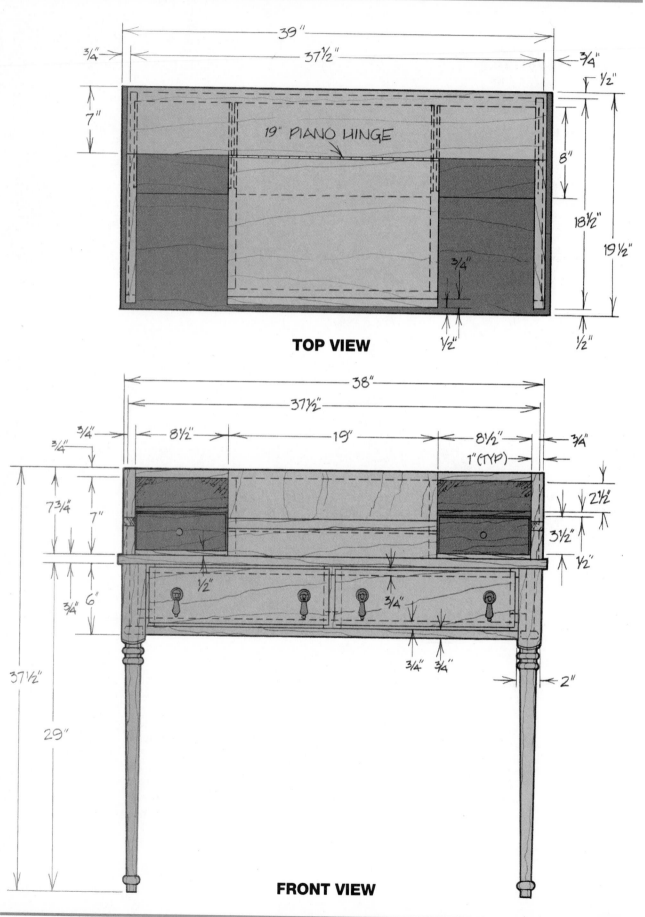

TOP VIEW

FRONT VIEW

2 Cut the joinery in the legs and aprons.

The aprons are joined to the legs by mortises and tenons. Cut these joints *before* you turn the legs. It's much easier to make a well-fitted joint while the stock is still square.

Cut the tenons in the ends of the aprons and the end stiles first. Set up a table saw and dado cutter to make 1⅛″-wide, ¼″-deep rabbets. Cut a rabbet in the end of each board, turn it over, and cut another. The two rabbets will form a tenon. Switch to an ordinary saw blade and miter the end of each tenon at 45°. Finally, cut ¾″-wide, 1⅛″-long notches in the top and bottom of the tenon with a band saw or a hand saw. (See Figures 1 through 3.)

1/Cut the tenons in the ends of the aprons with a dado cutter. Attach a stop block to the rip fence to automatically gauge the length of each tenon.

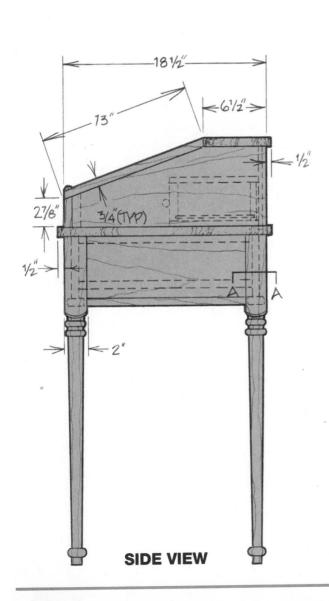

SIDE VIEW

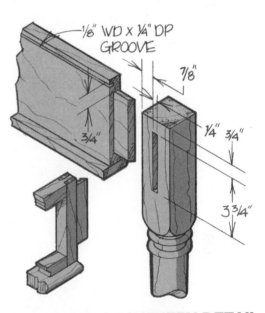

LEG-TO-APRON JOINERY DETAIL

⅛″ WD x ¼″ DP GROOVE

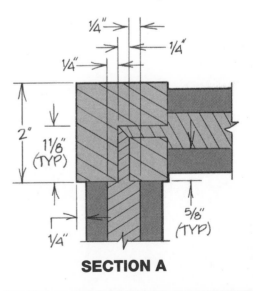

SECTION A

After cutting the tenons, make the mortises in the legs. Carefully mark the mortises on two adjacent faces of each leg. Using a drill press, rough out each mortise by drilling a series of ¼″-diameter, 1⅛″-deep holes.

Square the edges of the rough mortises with a chisel. (See Figures 4 and 5.) Each set of mortises should meet inside the leg stock, as shown in *Section A*.

2/Miter the end of each tenon at 45°, so the tenons will fit flush, end to end, in the mortises. Once again, use a stop block to automatically position the tenon over the saw blade.

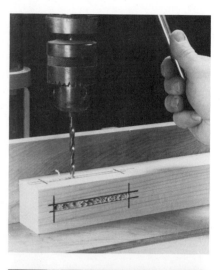

4/To rough out each mortise, drill a series of stopped holes with a drill press. Keep the line of holes perfectly straight by guiding the stock along a fence or straightedge.

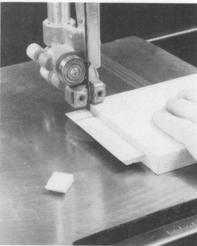

3/Notch the top and the bottom of each tenon with a band saw or hand saw, creating shoulders on the top and bottom. These shoulders make the mortise-and-tenon joint stronger and more stable.

5/Clean up the edges of each mortise with a chisel. Check the fit of the tenons in the mortise as you work. The fit should be snug, but not too tight.

3 **Turn the shape of the legs.** Turn each of the legs on a lathe, trying to match them as closely as possible. If you have a lathe duplicating jig, use it to shape the legs. If not, follow the procedure in Step-by-Step: Making Duplicate Turnings. Finish sand the legs on the lathe, and cut them to length.

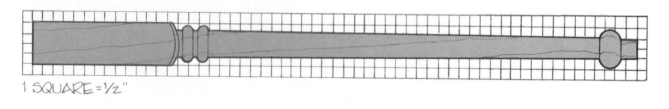

1 SQUARE = ½″

LEG LAYOUT

4

Make the front frame. The front apron is actually a frame, assembled with bridle joints. To make these joints, first cut slots in the ends of each stile, as shown in the *Front Frame Joinery Detail*. Then cut tenons in the ends of each rail. To cut the center joint, thin out an area in the rail by cutting two 1½″-wide, ¼″-deep notches on opposite sides of the rail. (See Figures 6 and 7.) Assemble the front frame with glue, checking that the frame is absolutely square as you clamp it together. When the glue dries, sand the joints clean and flush.

6/Cut slots in the stiles and tenons on the rails by guiding the stock over the saw blade with a tenoning jig. The jig shown can be easily made from scraps of wood. It rides along a table saw fence, guiding the wood over the blade.

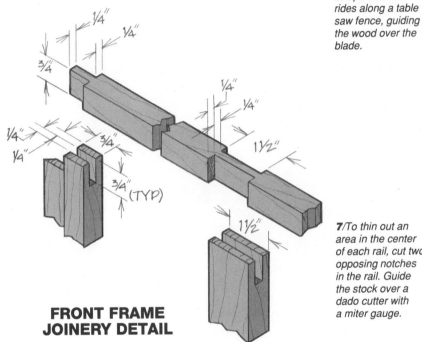

FRONT FRAME JOINERY DETAIL

7/To thin out an area in the center of each rail, cut two opposing notches in the rail. Guide the stock over a dado cutter with a miter gauge.

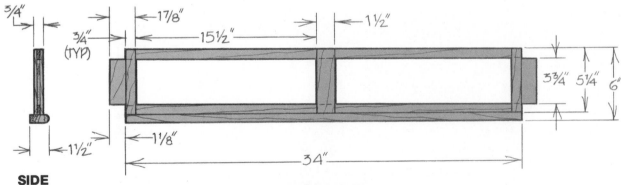

SIDE VIEW

FRONT VIEW

FRONT FRAME LAYOUT

5

Make the web frame. The drawers are supported by a web frame inside the desk. Assemble this frame with tongue-and-groove joinery. With a dado cutter or a router, cut ¼"-wide, ⅜"-deep grooves in the inside edges of the rails. Then cut ¼"-wide, ⅜"-long tenons (or tongues) in the ends of the stiles.

(See Figures 8 and 9.) Glue the rails and stiles together, as shown in the *Web Frame/Top View*. Check that the assembly is square as you clamp the parts together.

After the glue dries, cut notches in each of the frame corners with a hand saw or saber saw.

8/Cut a groove down the inside edges of the web frame rails. Use a fence to guide the stock, and a featherboard to help hold it against the fence.

9/Cut the tongues in the ends of the web frame stiles with the same tool you used to cut the grooves. Use the miter gauge to guide the stock, and a stop block to gauge the length of the tongues.

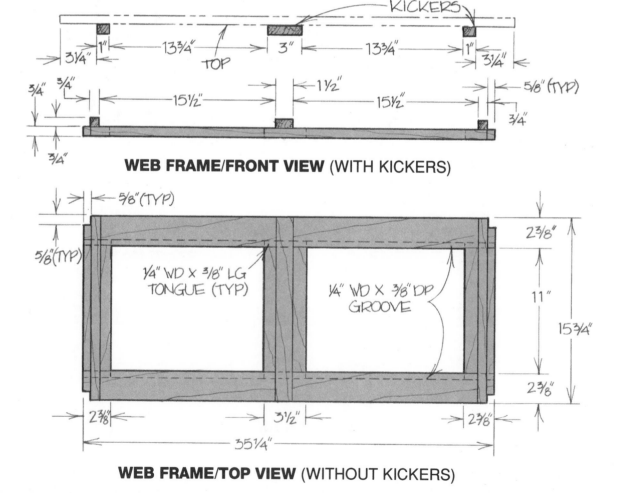

WEB FRAME/FRONT VIEW (WITH KICKERS)

WEB FRAME/TOP VIEW (WITHOUT KICKERS)

6 **Cut grooves in the aprons.** The desk top attaches to the aprons with clips, as shown in the *Desktop Joinery Detail*. Using a saw blade that cuts a ⅛″-wide kerf, cut a ⅛″-wide, ¼″-deep groove in the *inside* face of each apron, approximately ¼″ from the top edge. The precise distance from the top edge depends on the tabletop clips you're using.

7 **Assemble the table section.** Cut the molding that attaches to the bottom edge of the aprons and face frame. Using a router and a ¼″ quarter-round bit, round over the front edge of the molding, as shown in the *Apron Molding Joinery Detail*.

Finish sand all the parts and assemblies you've made so far, except the web frame. If you have a plate joiner, attach the molding to the aprons and front frame with #20 plates or biscuits. Use splines if you don't have a plate joiner. Assemble the legs, aprons, and front frame with glue. Rest the web frame on the moldings, inside the table assembly, as you clamp the parts together. This will hold the assembly square as the glue dries.

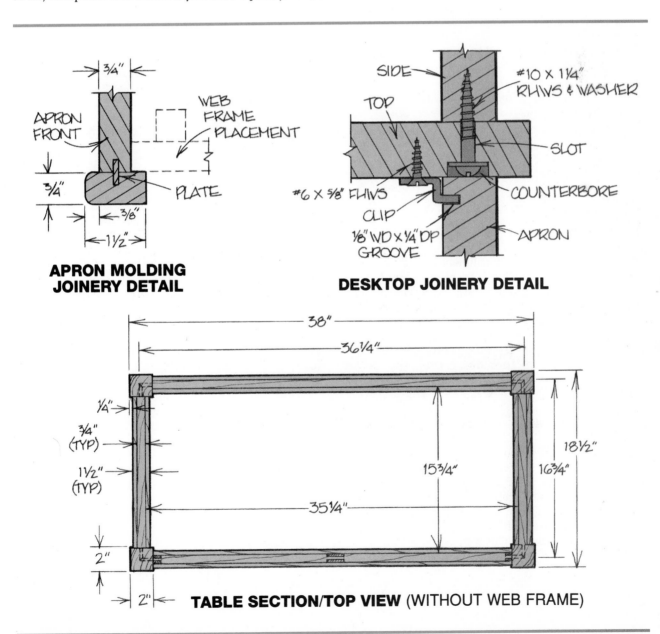

**APRON MOLDING
JOINERY DETAIL**

DESKTOP JOINERY DETAIL

TABLE SECTION/TOP VIEW (WITHOUT WEB FRAME)

Remove the web frame and position the desk top on the table assembly when the glue dries. Fasten the desk top to the apron with clips and screws. (It will be much easier to install the clips without the web frame in the way.) Remove the clips and set the desk top aside. Glue the web frame to the moldings and aprons, then glue the drawer guides to the web frame.

8 Cut the joinery in the sides, dividers, and back of the top section.
The shelves rest in ½"-wide, ¼"-deep blind dadoes and rabbets in the sides and dividers. Mark the rabbets and dadoes on the stock, as shown in the *Side/Divider Layout*. Rout these joints with a router and a straight bit, using a straightedge to guide the router. Square the blind ends with a chisel. Also, cut ¾"-wide, ⅜"-deep dadoes in the back to hold the dividers.

9 Cut the shapes of the sides, dividers, and shelves.
Mark the sloping top edge of the sides and dividers, and the notches in the front corners of the shelves. Cut these with a band saw or saber saw. Sand the sawed edges of the tapers.

> **TRY THIS!** Stack the sides and dividers and tape them together. Cut and sand all four parts at once. You can do the same for the short shelves. This saves time and ensures that the tapers and notches all match precisely.

10 Assemble the top section.
Round over all edges and ends of the desk top, the back edge and both ends of the top, and both edges of the side moldings and lid molding. Leave one corner of the moldings square, as shown in the *Side Molding Joinery Detail* and *Lid Molding Joinery Detail*. Finish sand the parts of the top section.

Cut slots for #20 plates to join the sides, back, and top. (As mentioned before, you can also use splines.) Dry assemble the sides, back, dividers, top, and shelves to check the fit of the parts, then assemble them with glue and plates. After the glue dries, sand the joints clean and flush.

Place the top section on the desk top and mark its position. Attaching the top section to the desk top presents a problem. If you glue the parts together, the desk top will not be able to expand and contract with changes in humidity. The grain direction of the sides and dividers is perpendicular to the grain direction of the desk top, and this will restrict its movement. Eventually, the desk top will split or warp.

Prevent this by screwing the top section to the desk top through *slots*. To make these slots, first drill ⅝"-diameter, ¼"-deep holes in the underside of the desk top, wherever you want to drive a screw. Inside these stopped holes, drill two or three ³⁄₁₆"-diameter holes in a straight line, perpendicular to the grain direction. These should form a slot approximately ³⁄₁₆" wide and ½" long, as shown in the *Counterbore and Slot Detail*. Drive #10 x 1¼" roundhead screws with flat washers through the slots and into the top assembly, as shown in the *Desk Top Joinery Detail*. Tighten the screws so they are snug, but not so tight that they will restrict the movement of the top section.

Attach the kickers to the underside of the desk top in the same manner. Drill counterbored slots in each kicker, then attach it to the top with roundhead wood screws and washers. When the top assembly and the kickers are secure, attach the desk top to the table assembly with the clips that you installed earlier.

Cut the side moldings, mitering the adjoining ends. Attach the moldings to the sides with glue and #20 plates (or splines). Also, attach the lid molding to the front end of the lid. Wait for the glue to dry and sand the joints clean and flush. Then attach the lid assembly to the top with a piano hinge.

11

Make and fit the drawers. Both the small and the large drawers are made in a similar manner: The sides are attached to rabbets in the drawer fronts, and the backs fit into dadoes in the sides. The bottoms float in grooves cut into the other drawer parts. The only difference is that the large drawers have a lip all around the drawer front, while the small drawers have none. Here's a list of the joints and parts to cut:

- ⅞"-wide, ⅜"-deep rabbets in the ends of the large drawer fronts

- ½"-wide, ⅝"-deep rabbets in the ends of the small drawer fronts
- ½"-wide, ¼"-deep dadoes in both the large and small drawer sides, ¼" from the back ends
- ⅜"-wide, ⅜"-deep rabbets in the top and bottom edges of the large drawer fronts
- ¼"-wide, ¼"-deep grooves in the inside faces of all the drawer parts, ¼" from the bottom edge

Round over the edges of the large drawer fronts, using a router and a ¼" quarter-round bit. Drill the fronts for drawer pulls, and finish sand the outside surfaces. It's not necessary to sand the other drawer parts.

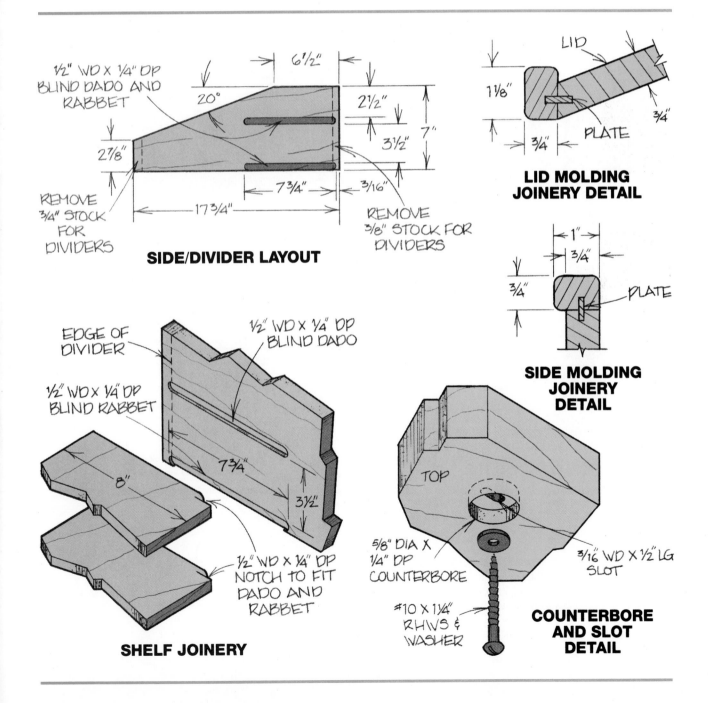

SIDE/DIVIDER LAYOUT

LID MOLDING JOINERY DETAIL

SIDE MOLDING JOINERY DETAIL

SHELF JOINERY

COUNTERBORE AND SLOT DETAIL

Assemble the drawer fronts, sides, and backs with glue. Slide the bottoms into their grooves as you assemble the other parts, but *don't* glue them in place. Let them float in the grooves. Wait for the glue to set, then reinforce the rabbet and dado joints with finishing nails. Drive the nails through the sides and into the drawer fronts or backs. Set the heads of the nails.

Install pulls in the drawer fronts, and fit the drawers in the desk. Test the sliding action. If any drawer sticks, remove a little stock from it with a file or a scraper.

TRY THIS! Many cabinetmakers prefer to make drawers ¹⁄₁₆″–⅛″ oversize, then sand or scrape them to fit. This way, they can fit each drawer perfectly.

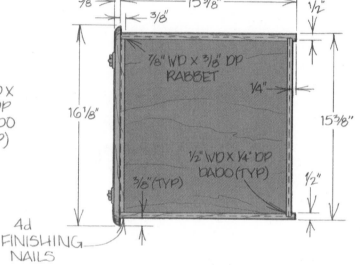

TOP VIEW

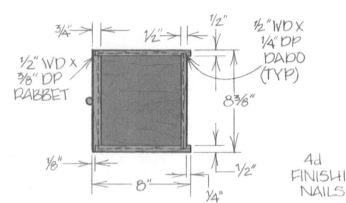

TOP VIEW

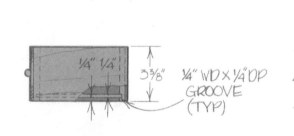

SIDE VIEW

SMALL DRAWER

SIDE VIEW

LARGE DRAWER

12 Finish the completed desk. Remove the drawers and the lid from the desk, then detach the top assembly from the table assembly. Remove all the hardware and set it aside. Do any necessary touch-up sanding and apply a finish. Coat all wooden surfaces, except for the insides of the drawers, the kickers, and the drawer guides. (Traditionally, only the fronts of drawers are finished. If you apply finish to other drawer parts, it may interfere with the sliding action.) Be careful to coat the top and bottom of the desk top evenly — this will prevent it from warping. When the finish dries, wax and buff the surfaces that will show. Reassemble the desk.

Credits

Contributing Craftsmen and Craftswomen:

Larry Callahan (Schoolmarm's Desk)

Judy Ditmer (Desk Carousel)

Nick Engler (Child's Rolltop Desk, Wooden Filing
Cabinet, Multipurpose Taboret, Hanging Bookcase,
Knife Box Diskette File)

W. R. Goehring (Contemporary/Classic Writing Table)

Jim McCann (Computer Workstation)

Virgil Schmidt (Drafting Table)

Note: Several of the projects in this book were built by
craftsmen or craftswomen whose names have been
erased by time. We regret that we cannot tell you
who built them; we can only admire their craftsman-
ship. These pieces include the Adjustable Bookcase
and Cabinet, Barrister's Bookcase, and Letter Holder.

The Computer Workstation is copyrighted by Shopsmith,
Inc. The designs for the remainder of the projects in this
book (those attributed to a designer/builder) are the
copyrighted property of the craftsmen and craftswomen
who made them. Readers are encouraged to reproduce
these copyrighted projects for their personal use or for
gifts. However, reproduction for sale or profit is forbid-
den by law.

Special Thanks To:
 Mr. and Mrs. Douglas Crowell
 Mr. and Mrs. Nicholas Engler, Jr.
 Gordon Honeyman
 Heartwood, Tipp City, Ohio
 Shopsmith, Inc., Dayton, Ohio
 Wertz Hardware Store, West Milton, Ohio

Rodale Press, Inc., publishes AMERICAN WOODWORKER™, the magazine for the serious woodworking hobbyist. For information on how to order your subscription, write to AMERICAN WOODWORKER™, Emmaus, PA 18098.

WOODWORKING GLOSSARY

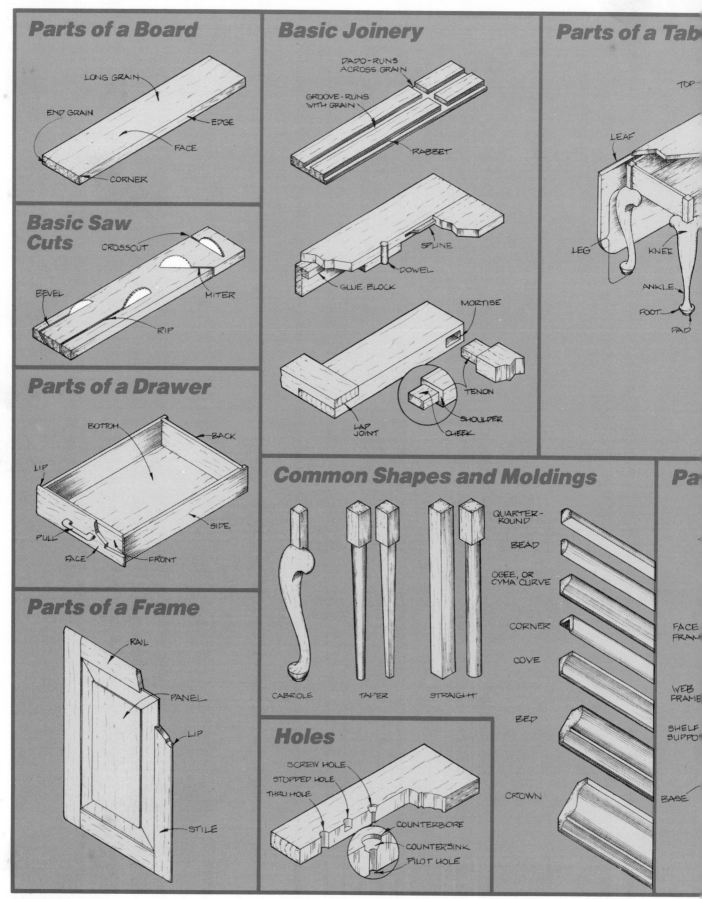

Parts of a Board

LONG GRAIN
END GRAIN
EDGE
FACE
CORNER

Basic Saw Cuts

CROSSCUT
BEVEL
MITER
RIP

Parts of a Drawer

BOTTOM
BACK
LIP
SIDE
PULL
FACE
FRONT

Parts of a Frame

RAIL
PANEL
LIP
STILE

Basic Joinery

DADO - RUNS ACROSS GRAIN
GROOVE - RUNS WITH GRAIN
RABBET
SPLINE
DOWEL
GLUE BLOCK
MORTISE
TENON
SHOULDER
CHEEK
LAP JOINT

Parts of a Tab[le]

TOP
LEAF
LEG
KNEE
ANKLE
FOOT
PAD

Common Shapes and Moldings

QUARTER-ROUND
BEAD
OGEE, OR CYMA CURVE
CORNER
COVE
BED
CROWN
CABRIOLE
TAPER
STRAIGHT

Holes

SCREW HOLE
STOPPED HOLE
THRU HOLE
COUNTERBORE
COUNTERSINK
PILOT HOLE

Pa[rts]

FACE FRAM[E]
WEB FRAME
SHELF SUPPO[RT]
BASE